AF878617

EAI/Springer Innovations in Communication and Computing

Series editor:

Imrich Chlamtac, CreateNet, Trento, Italy

Editor's Note

The impact of information technologies is creating a new world yet not fully understood. The extent and speed of economic, life style and social changes already perceived in everyday life is hard to estimate without understanding the technological driving forces behind it. This series presents contributed volumes featuring the latest research and development in the various information engineering technologies that play a key role in this process.

The range of topics, focusing primarily on communications and computing engineering include, but are not limited to, wireless networks; mobile communication; design and learning; gaming; interaction; e-health and pervasive healthcare; energy management; smart grids; internet of things; cognitive radio networks; computation; cloud computing; ubiquitous connectivity, and in mode general smart living, smart cities, Internet of Things and more. The series publishes a combination of expanded papers selected from hosted and sponsored European Alliance for Innovation (EAI) conferences that present cutting edge, global research as well as provide new perspectives on traditional related engineering fields. This content, complemented with open calls for contribution of book titles and individual chapters, together maintain Springer's and EAI's high standards of academic excellence. The audience for the books consists of researchers, industry professionals, advanced level students as well as practitioners in related fields of activity include information and communication specialists, security experts, economists, urban planners, doctors, and in general representatives in all those walks of life affected ad contributing to the information revolution.

About EAI

EAI is a grassroots member organization initiated through cooperation between businesses, public, private and government organizations to address the global challenges of Europe's future competitiveness and link the European Research community with its counterparts around the globe. EAI reaches out to hundreds of thousands of individual subscribers on all continents and collaborates with an institutional member base including Fortune 500 companies, government organizations, and educational institutions, provide a free research and innovation platform.

Through its open free membership model EAI promotes a new research and innovation culture based on collaboration, connectivity and recognition of excellence by community.

More information about this series at http://www.springer.com/series/15427

EAI

RESEARCH MEETS INNOVATION

Maroš Finka • Matej Jaššo • Milan Husár
Editors

The Role of Public Sector in Local Economic and Territorial Development

Innovation in Central, Eastern and South Eastern Europe

Springer

Editors
Maroš Finka
Spectra Centre of Excellence of the EU
Slovak University of Technology in
Bratislava
Bratislava, Slovakia

Matej Jaššo
Spectra Centre of Excellence of the EU
Slovak University of Technology in
Bratislava
Bratislava, Slovakia

Milan Husár
Spectra Centre of Excellence of the EU
Slovak University of Technology in
Bratislava
Bratislava, Slovakia

ISSN 2522-8595 ISSN 2522-8609 (electronic)
EAI/Springer Innovations in Communication and Computing
ISBN 978-3-319-93574-4 ISBN 978-3-319-93575-1 (eBook)
https://doi.org/10.1007/978-3-319-93575-1

Library of Congress Control Number: 2018951798

© Springer International Publishing AG, part of Springer Nature 2019
This work is subject to copyright. All rights are reserved by the Publisher, whether the whole or part of the material is concerned, specifically the rights of translation, reprinting, reuse of illustrations, recitation, broadcasting, reproduction on microfilms or in any other physical way, and transmission or information storage and retrieval, electronic adaptation, computer software, or by similar or dissimilar methodology now known or hereafter developed.
The use of general descriptive names, registered names, trademarks, service marks, etc. in this publication does not imply, even in the absence of a specific statement, that such names are exempt from the relevant protective laws and regulations and therefore free for general use.
The publisher, the authors, and the editors are safe to assume that the advice and information in this book are believed to be true and accurate at the date of publication. Neither the publisher nor the authors or the editors give a warranty, express or implied, with respect to the material contained herein or for any errors or omissions that may have been made. The publisher remains neutral with regard to jurisdictional claims in published maps and institutional affiliations.

This Springer imprint is published by the registered company Springer Nature Switzerland AG
The registered company address is: Gewerbestrasse 11, 6330 Cham, Switzerland

Preface

This book reflects current outputs from the research of the member institutions of the Network of Spatial Research and Planning in Central, Eastern and South Eastern Europe—SPA-CE.net and their contribution to professional debate on the role of the public sector in innovative local economic and territorial development in Central, Eastern and South Eastern Europe. The ambition of this book, published in collaboration with the European Alliance for Innovation, is to foster broader discussion on challenges and changes in the position of the public sector in the development and spread of innovations across different sectors. The SPA-CE.net follows with this book its mission to ensure sustainable development on the European continent by providing spatial research results to policymakers, scientists and the interested public, to contribute significantly to improving the quality of life, to enhancing the protection and management of Europe's cultural and natural heritage as well as creating new jobs and spurring economic growth and to mobilising Europe's potential for research and practical application in the field of sustainable spatial development.

Bratislava, Slovakia

Maroš Finka
Matej Jaššo
Milan Husár

Acknowledgements

This book has been published thanks to the support of the SPA-CE.net, European Alliance for Innovation in Bratislava and Central European Research and Training Centre in Spatial Planning—SPECTRA Centre of Excellence of the European Union at the Slovak University of Technology in Bratislava. The initiating event for the work on this publication was supported by the University of Architecture, Civil Engineering and Geodesy in Sofia, Leibniz Institute of Ecological Urban and Regional Development (IOER) and SPECTRA CE EU at the STU in Bratislava.

Acknowledgements

[illegible]

Introduction

Support of sustainable communities' development at the local and regional level helping people to shape better communities through participation, education, information and the effective use of technology belongs recently to the main tasks for public authorities. Community sustainability is directly linked to the sustainable territorial development and the development of local economies that are socially responsible, economically viable and environmentally sound. To face responsibly this task requires the participation of the whole public sector to identify community needs and potentials, to find appropriate interventions and to implement innovative solutions.

In the contexts of current challenges and problems in territorial development in Central, Eastern and South Eastern Europe, smart sustainable growth in all spheres of community life seems to be the target quality in territorial as well as economic development. Civil society development, forthcoming European integration, transformation towards knowledge-based economy, demographic shifts, strengthened environmental awareness, increased fiscal concerns and more nuanced views of growth determine the framework not only for territorial and economic development itself, but for its management and for the role of the public sector in it as well. The result is the formulation of new qualitative and quantitative demands and new opportunities for smart growth.

Although the features distinguishing smart growth in a community can vary from place to place and from sphere to sphere of community life, in general, smart growth means to invest time, creativity and resources in innovative solutions, restoring community and developing vital local and regional economy delivering sustainable working places and meeting available labour market potential.

This book shows the potential of the public sector in the Central, South and South Eastern European countries represented by national, regional and local governmental and self-governmental bodies, public institutions, non-governmental organisations, groups of activists and professionals to help communities to identify and leverage community assets to create sustainable policies, plans and practices reflecting dynamic development and unique framework preconditions and challenges in new innovative planning and development concepts.

Chapter 2 "Innovations and changing role of public sector in spatial development strategies" discusses the problems and challenges for local and regional development in Central and Eastern European Countries resulting from the global transformation processes of the society and economy accelerated by the transition process in the majority of the countries in Central and Eastern Europe. The authors address societal and economic transformation processes framing the innovations and changing the role of the public sector in spatial development strategies, challenges for economic and territorial development relevant for new definition of the role of the public sector related to the development as the object of the policy and related to the process of the governance and decision-making. The chapter shows that the identified deep changes and principal challenges need proper response in the content and in the processes of the elaboration of spatial development strategies as integrative and integrated strategies as well as particular strategies for the development of spatial relevant fields of societal life. Conceptual innovations have to be followed by the substantial, instrumental and processual innovations reaching from institutional via behavioural up to technological as it is shown in the following chapters.

Chapter 3 "Spatial planning policies and the integration models as a means for a better delivery of services of general interest" is devoted to new approaches of the services of general interest (SGI) provision, especially in remote mountain territories and border areas. Harsh economic conditions, population ageing, dispersed settlement and the geomorphology hinder the provision and supply of the SGI in these territories. In order to overcome this deficiency, new approaches of the SGI provision should be sought and identified. Therefore, spatial planning and sectorial policies and the related models of integration (sectors, actors, funds, services, policies, etc.) were investigated. The analysis of the coverage of the SGI in the spatial planning and other (space related) sectorial policies documents (regulations, strategies, plans, guidelines, etc.) was done in five Alpine countries (Italy, Switzerland, Austria, France and Slovenia). Similarities and distinctions among those five countries are entirely discussed.

Chapter 4 "Utilising endogenous potentials via regional policy-led development initiatives in (post-) industrial regions of Central Europe" analyses the ways how place-based endogenous potentials are conceptualised and used, particularly in the local and regional context. Places characterised as non-agglomeration, (post-) industrial regions in Central Europe, often situated in the spatial peripheries, are in the spotlight. This contribution highlights governance issues and the valorisation of endogenous development potentials by different policy levels, discussing agenda setting and implementation. The authors argue that utilising potentials is streamlined top-down via various policy frameworks, which are in turn mediated by local and regional governance settings, adapting and translating these programmes into practical, regional and local actions, and thereby creating important learning effects for the regions.

Chapter 5 "Energy sensitive spatial planning as a public sector tool towards sustainable economic and territorial development" is focused on sustainable energy transition, which is increasingly regarded as a key topic of both policy and research in the EU context of the global process towards sustainable development. By linking diverse sociocultural contexts and setting a particular focus on the local governance level, the chapter discusses current results of ongoing ECHOES project (HORIZON 2020) with regard to the literature review on energy planning and management and the discussions undertaken within three preliminary focus groups in Bulgaria. The authors then relate them to the findings of an accomplished university research project in the University of Architecture, Civil Engineering and Geodesy (UACEG) in Sofia, aimed at developing a framework for evaluation of the relevance and effectiveness of strategic planning documents applied at the local/municipal level in Bulgaria to sustainable development requirements.

Chapter 6 “Determining land values by location: supporting public valuation expert committees in the provision of market transparency” is focused on land values which reflect the spatial location within a territory. Different criteria like economic situation or social condition are discussed. A consistent method of determining land values nationwide for all locations increases the provision of a high market transparency. Comprehensive decision-making basis for all market participants as well as for the public sector is a precondition for effective management of spatial development. The practical implication of the concept is shown in the case of Germany.

Chapter 7 “Forest commons as a model for territorial governance” refers to the territorial governance patterns generating perspective, place-based and highly adaptable models. Most territorial governance studies take a socio-institutional or spatial planning perspective. Still, territorial governance happens in and through other policy domains—this chapter examines ecosystem-based forest commons’ governance. Resilience of forest commons as resources and as institutions leads the society towards sustainable territorial development at local and global scales. The illustration of this account comes through exploring the Albanian model of forest commons within a river basin. The model embodies ecosystem values besides utilitarian ones and discusses the dimensions of territorial governance for forest commons. Evidence is collected through visual surveys, interviews and focus groups systematically researching a number of commons’ variables of endurance and functionality.

Chapter 8 “The role of leadership in the current contexts of Central European planning culture” is devoted to the relation between two of the most significant soft factors in spatial development, namely leadership and planning culture. The roles of planners have been recently dramatically modified—from the position of the designers of the future towards the mediators and leaders of the development processes. Understanding of cultural dynamics, the ability to develop a clear vision of the desired planning culture based on the democratic leadership and its active fostering should be an essential competence of any planner. It has been shown that highly profiled planning cultures based on the active leadership generating typical set of principles, values and approaches play an important role in the spatial development of any territory. This chapter introduces fundamental theoretical and methodological models of planning cultures and tries to investigate and define specific features and characteristics of Central European planning culture, including the field of leadership.

Chapter 9 “Urban transformation - cultural heritage and local economic development: the public sector on the battlefront?” investigates the phenomena of historical cities in Central and Eastern Europe beyond the booming metropolis, facing stagnation or decline of population and economic development. These cities might regain their competitiveness, but the existing building stock does not often meet the needs of housing or retail. Peculiarities of coordination process between the public and the private sector as well as within the public sector necessary to solve the conflict between cultural heritage preservation and local economic development are examined in the case of city of Görlitz. This chapter reflects the first experiences from matrix assessing balance between heritage preservation and urban development.

“Social innovation and sustainable economic development: participatory tourism destination management” is the main scope of chapter 10. This chapter addresses the role of bottom-up innovation to support sustainable local development, in particular in local tourism development, and introduces the case study of Gramsh, a remote Albanian region. This case shows that sustainable tourism can be developed in peripheral territories with typical problems like high rate of unemployment, lack of infrastructure, emigration, lack of investments and rather weak local government. Bottom-up driven tourism innovation strengthens the endogenous development potential. Such an approach fosters local identity formation and self-identification of the citizens with their own territory and local resources. Gramsh case study is a valuable contribution to searching for sustainable tourism management models, not only in the Western Balkans.

Chapter 11 deals with “Professional capacity building as the public sector intervention towards sustainable economic and territorial development” and introduces a key study devoted to knowledge hubs network for energy efficiency. This chapter is focused on the public sector role in professional capacity building necessary for the improvement of energy efficiency in the building sector at the national and local level, which has direct implications for the territorial and urban planning and the design of the national strategic documents and programmes targeted to the residential buildings. It brings together the results achieved by four consecutive EU-financed projects in reforming of the outdated training practices and implementation of nearly zero-energy building (NZEB) in Bulgaria and, consequently, in other countries in Eastern and Southern Europe.

Chapter 12 “The public in search of identity: new symbolism in urban spaces. A study of central squares of Balkan capitals” is devoted to thorough analysis of the process of building new political and cultural identities in the Western Balkans. This chapter presents a study of identity formation, specifically how identity in the four Western Balkan countries is expressed in significant place. The alterations of central squares of capitals, regarded as tangible expressions of national identity, have been studied in four small countries: Albania, Kosovo, Macedonia and Montenegro. The study focuses on how selective remembrance and active forgetting played a role in the development of main squares, how ambiguous spaces have been created with also banal practices and how the squares can be seen in view of globalised identities and cultural hybridities. Cases of cities of Tirana, Pristina, Podgorica and Skopje illustrate this process.

Last chapter "Professional associations as public actors in the formulation and implementation of spatial development policies. Key study monitoring and evaluation standard for the urban environment of Sofia" deals with the gradual shifts of the role of planning profession, its institutions and planning in general in Bulgaria. It introduces the conceptual context of the transactive planning in spatial development and lower level institutional thickness. Thorough analysis of the key studies of the Association of Bulgarian Urban and Regional Planners (ABURP) demonstration projects "Share the Neighbourhood" and "Set an Urban Standard" in Sofia is the main scope of this chapter.

Spectra Centre of Excellence of the EU
Slovak University of Technology in Bratislava
Bratislava, Slovakia

Maroš Finka
Matej Jaššo
Milan Husár

Contents

Contributors

Besnik Aliaj Polis University, Tirana, Albania

Angel Burov Association of Bulgarian Urban and Regional Planners (ABURP), Sofia, Bulgaria

Aida Ciro Department of Territorial Governance and Development, Permanent Workshop on Territorial Governance/Co-PLAN, Institute for Habitat Development/POLIS University, Tirana, Albania

Elena Dimitrova Department of Urban Planning, University of Architecture, Civil Engineering and Geodesy, Sofia, Bulgaria

Maroš Finka Spectra Centre of Excellence of the EU, Slovak University of Technology in Bratislava, Bratislava, Slovakia

Jörn Harfst Department of Geography and Regional Science, University of Graz, Graz, Austria

Milan Husár Spectra Centre of Excellence of the EU, Slovak University of Technology in Bratislava, Bratislava, Slovakia

Ľubomír Jamečný Spectra Centre of Excellence of the EU, Slovak University of Technology in Bratislava, Bratislava, Slovakia

Matej Jaššo Spectra Centre of Excellence of the EU, Slovak University of Technology in Bratislava, Bratislava, Slovakia

Robert Knippschild Leibniz Institute of Ecological Urban and Regional Development & Technische Universität Dresden, Dresden, Germany

Špela Kolarič Department of Landscape Architecture, Biotechnical Faculty, University of Ljubljana, Ljubljana, Slovenia

Vasil Madzihirski Association of Bulgarian Urban and Regional Planners (ABURP), Sofia, Bulgaria

Barbara Černič Mali Department of Landscape Architecture, Biotechnical Faculty, University of Ljubljana, Ljubljana, Slovenia

Naja Marot Department of Landscape Architecture, Biotechnical Faculty, University of Ljubljana, Ljubljana, Slovenia

Irina Mutafchiiska Association of Bulgarian Urban and Regional Planners (ABURP), Sofia, Bulgaria

Peter Nientied Co-Plan Institute for Habitat Development, POLIS University, Tirana, Albania

Institute for Housing and Urban Development Studies, Erasmus University, Rotterdam, Netherlands

Vladimír Ondrejička Spectra Centre of Excellence of the EU, Slovak University of Technology in Bratislava, Bratislava, Slovakia

Andreas Ortner Chair of Land Management, TU Dresden, Dresden, Germany

Danko Simić Department of Geography and Regional Science, University of Graz, Graz, Austria

Matthias Soot Chair of Land Management, TU Dresden, Dresden, Germany

Milena Tasheva-Petrova Department of Urban Planning, University of Architecture, Civil Engineering and Geodesy, Sofia, Bulgaria

Merita Toska Department of Urban and Regional Governance and Development, Permanent Workshop on Territorial Governance/Co-PLAN, Institute for Habitat Development/POLIS University, Tirana, Albania

Rudina Toto Permanent Workshop on Territorial Governance, Co-PLAN, Institute for Habitat Development, POLIS University, Tirana, Albania

Dragomir Tzanev EnEffect, Center for Energy Efficiency, Sofia, Bulgaria

Alexandra Weitkamp Chair of Land Management, TU Dresden, Dresden, Germany

Peter Wirth Leibniz Institute of Ecological Urban and Regional Development (IOER), Dresden, Germany

Constanze Zöllter Leibniz Institute of Ecological Urban and Regional Development, Dresden, Saxony, Germany

Abbreviations

ABURP	Association of Bulgarian Urban and Regional Planners
ARL	Akademie für Raumforschung und Landesplanung
ASUB	Association of Students of Urbanism in Bulgaria
AT	Austria
BKH	Building Knowledge Hubs
BBSR	Das Bundesamt für Bauwesen und Raumordnung
BL	Building land
CEE	Central and Eastern European
CEMAT	Council of Europe of Ministers Responsible for Spatial Planning
CLLD	Community-Led Local Development
CPR	Common pool resources
DUI	Democratic Union for Integration
EE	Energy efficiency
EEA	European Environment Agency
EU	European Union
ERDF	European Regional Development Fund
ESDP	European Spatial Development Planning Network
ESF	European Social Fund
FYR	Former Yugoslav Republic
FRA	France
GIS	Geographic Information Systems
GLTP	General Local Territorial Plan
GSPM	General Spatial Plans of the municipalities
IC	Information communication
ICT	Information communication technology
IT	Italy
ITI	Integrated Territorial Investments
IOER	Institute of Ecological Urban and Regional Development
IPURD	Integrated Plans for Urban REgeneration and Development
ITI	Integrated Territorial Investments
LEADER	Liaison Entre Actions de Développement de l'Économie Rurale

MDP	Municipal development plans
MVH	Mountain Virtual Hospital
NAVET	National Agency for Vocational Education and Training
NGO	Non-government organisation
NSGI	Non-economic services
NUTS	Nomenclature des Unitées Territoriales Statistiques
NZEB	Nearly zero-energy building
OECD	Organisation for Economic Co-operation and Development
PH	Passive house
PHI	Passive house Institute
RENEUER	Regional Network for Efficient Use of Energy and Water Resources
RES	Renewable energy resources
RCCS	Regional Card for Services
RD	Regional development
SD	Sustainable development
SDAASP	Departmental Plan of Improved Accessibility to Services of General Interest
SEA	Strategic Environment Assessments
SEE	South Eastern Europe
SET Plan	Strategic Energy Technology Plan
SES	Socio-ecological systems
SESMAD	Socio-Ecological Systems Meta-Analysis Database
SGEI	Services of general economic interest
SGI	Services of general interest
SWOT	Strengths, weaknesses, opportunities, threats
TAR	Territorial Administrative Reform
UACEG	University of Architecture, Civil Engineering and Geodesy
UBL	Untapped building land
UK	United Kingdom
UNMIK	United Nations Interim Administration Mission in Kosovo
WWII	World War II

Chapter 1
Editorial

Maroš Finka, Matej Jaššo, and Milan Husár

Support of sustainable communities' development at the local and regional level helping people to shape better communities through participation, education, information, and the effective use of technology belongs recently to the main tasks for public authorities. Community sustainability is directly linked to the sustainable territorial development and the development of local economies that are socially responsible, economically viable, and environmentally sound. To face responsibly this task requires the participation of the whole public sector to identify community needs and potentials, to find appropriate interventions and implement innovative solutions.

In the contexts of current challenges and problems in territorial development in Central, Eastern, and South Eastern Europe smart sustainable growth in all spheres of community life seems to be the target quality in territorial as well as economic development. Civil society development, forth coming European integration, transformation towards knowledge-based economy, demographic shifts, strengthened environmental awareness, increased fiscal concerns, and more nuanced views of growth determine the framework not only for territorial and economic development itself, but for its management and for the role of public sector in it as well. The result is the formulation of new qualitative and quantitative demands and new opportunities for smart growth.

Although the features distinguishing smart growth in a community can vary from place to place and from sphere to sphere of community life, in general, smart growth means to invest time, creativity, and resources in innovative solutions, restoring community and developing vital local and regional economy delivering sustainable working places and meeting available labour market potential.

M. Finka (✉) · M. Jaššo · M.Husár
Spectra Centre of Excellence of the EU, Slovak University of Technology in Bratislava, Bratislava, Slovakia
e-mail: maros.finka@stuba.sk; matej.jasso@stuba.sk; milan.husar@stuba.sk

© Springer International Publishing AG, part of Springer Nature 2019
M. Finka et al. (eds.), *The Role of Public Sector in Local Economic and Territorial Development*, EAI/Springer Innovations in Communication and Computing, https://doi.org/10.1007/978-3-319-93575-1_1

This book shows the potential of the public sector in the Central, South, and South Eastern European countries represented by national, regional, and local governmental and self-governmental bodies, public institutions, non-governmental organisations, groups of activists, and professionals to help communities to identify and leverage community assets to create sustainable policies, plans, and practices reflecting dynamic development and unique framework preconditions and challenges in new innovative planning and development concepts.

Chapter 2 "Innovations and changing role of public sector in spatial development strategies" discusses the problems and challenges for local and regional development in Central and Eastern European Countries resulting from the global transformation processes of the society and economy accelerated by the transition process in majority of the countries in Central and Eastern Europe. The authors are addressing societal and economic transformation processes framing the innovations and changing the role of public sector in spatial development strategies, challenges for economic and territorial development relevant for new definition of the role of the public sector related to the development as the object of the policy and related to the process of the governance and decision-making. The chapter shows that the identified deep changes and principal challenges need proper responds in the content and in the processes of the elaboration of spatial development strategies as integrative and integrated strategies as well as particular strategies for the development of spatial relevant fields of societal life. Conceptual innovations have to be followed by the substantial, instrumental, and processual innovations reaching from institutional, via behavioural up to technological as it is shown in the following chapters.

Chapter 3 "Spatial planning policies and the integration models as a means for a better delivery of services of general interest" is devoted to new approaches of the services of general interest (SGI) provision, especially in remote mountain territories and border areas. Harsh economic conditions, population ageing, dispersed settlement, and the geomorphology hinder the provision and supply of the SGI in these territories. In order to overcome this deficiency, new approaches of the SGI provision should be sought and identified. Therefore, spatial planning and sectorial policies and the related models of integration (sectors, actors, funds, services, policies, etc.) were investigated. The analysis of the coverage of the SGI in the spatial planning, and other (space-related) sectorial policies documents (regulations, strategies, plans, guidelines, etc.) was done in five Alpine countries (Italy, Switzerland, Austria, France. and Slovenia). Similarities and distinctions among those five countries are entirely discussed.

Chapter 4 "Utilising endogenous potentials via regional policy-led development initiatives in (post-) industrial regions of Central Europe" analyses the ways, how place-based endogenous potentials are conceptualised and used, particularly in the local and regional context. Places characterised as non-agglomeration, (post-) industrial regions in Central Europe, often situated in the spatial peripheries are in the spotlight. This contribution highlights governance issues and the valorisation of endogenous development potentials by different policy levels, discussing agenda setting and implementation. The authors argue that utilising potentials is streamlined top-down via various policy frameworks, which are in turn mediated by local and regional governance settings, adapting and translating these programmes into practical, regional, and local actions, and thereby creating important learning effects for the regions.

Chapter 5 "Energy sensitive spatial planning as a public sector tool towards sustainable economic and territorial development" is focused on sustainable energy transition, which is increasingly regarded as a key topic of both—policy and research in the EU context of the global process towards sustainable development. By linking diverse sociocultural contexts and setting a particular focus on the local governance level, the chapter discusses current results of ongoing ECHOES project (HORIZON 2020) with regard to the literature review on energy planning and management, and the discussions undertaken within three preliminary focus groups in Bulgaria. The authors then relate them to the findings of an accomplished university research project in the University of Architecture, Civil Engineering and Geodesy (UACEG) in Sofia, aimed at developing a framework for evaluation of the relevance and effectiveness of strategic planning documents applied at the local/municipal level in Bulgaria to sustainable development requirements.

Chapter 6 "Determining land values by location: supporting public valuation expert committees in the provision of market transparency" is focused on land values which reflect the spatial location within a territory. Different criteria like economic situation or social condition are discussed. A consistent method of the determining land values nationwide for all locations increases the provision of a high market transparency. Comprehensive decision-making basis for all market participants as well as for the public sector is a precondition for effective management of spatial development. The practical implication of the concept is shown on the case of Germany.

Chapter 7 "Forest commons as a model for territorial governance" refers to the territorial governance patterns generating perspective, place-based, and highly adaptable models. Most territorial governance studies take a socio-institutional or spatial planning perspective. Still, territorial governance happens in and through other policy domains—this chapter examines ecosystem-based forest commons' governance. Resilience of forest commons as resources and as institutions, leads the society towards sustainable territorial development at local and global scales. The illustration of this account comes through exploring the Albanian model of forest commons within a river basin. The model embodies ecosystem values besides utilitarian ones and discusses the dimensions of territorial governance for forest commons. Evidences are collected through visual surveys, interviews, and focus groups systematically researching a number of commons' variables of endurance and functionality.

Chapter 8 "The role of leadership in the current contexts of Central European planning culture" is devoted to the relation between two of the most significant soft factors in spatial development, namely leadership and planning culture. The roles of planners have been recently dramatically modified—from the position of the designers of the future towards the mediators and leaders of the development processes. Understanding of cultural dynamics, the ability to develop a clear vision of the desired planning culture based on the democratic leadership and its active fostering, should be an essential competence of any planner. It has been shown that highly profiled planning cultures based on the active leadership generating typical set of principles, values, and approaches play an important role in the spatial development of any territory. This chapter introduces fundamental, theoretical, and methodological models of planning cultures and tries to investigate and define specific features and characteristics of Central European planning culture, including the field of leadership.

Chapter 9 "Urban transformation – cultural heritage and local economic development: the public sector on the battlefront?" investigates the phenomena of historical cities in Central and Eastern Europe beyond the booming metropoles, facing stagnation or decline of population and economic development. These cities might regain their competitiveness, but the existing building stock does often not meet the needs of housing or retail. Peculiarities of coordination process between the public and the private sector as well as within the public sector necessary to solve the conflict between cultural heritage preservation and local economic development are examined in the case of city of Goerlitz. This chapter reflects the first experiences from matrix assessing balance between heritage preservation and urban development.

"Social innovation and sustainable economic development: participatory tourism destination management" is the main scope of chapter 10. This chapter addresses the role of bottom up innovation to support sustainable local development, in particular in local tourism development and introduce the case study of Gramsh, a remote Albanian region. This case shows that sustainable tourism can be developed in peripheral territories with typical problems like high rate of unemployment, lack of infrastructure, emigration, lack of investments, and rather weak local government. Bottom up driven tourism innovation strengthens the endogenous development potential. Such an approach fosters local identity formation and self-identification of the citizens with their own territory and local resources. Gramsh case study is a valuable contribution to searching for sustainable tourism management models, not only in Western Balkan.

Chapter 11 deals with "Professional capacity building as the public sector intervention towards sustainable economic and territorial development" and introduces a key study devoted to knowledge hubs network for energy efficiency. This chapter is focused on the public sector role in professional capacity building necessary for the improvement of energy efficiency in the building sector at the national and local level, which has direct implications on the territorial and urban planning and the design of the national strategic documents and programmes targeted to the residential buildings. It brings together the results achieved by four consecutive EU-financed projects in reforming of the outdated training practices and implementation of nearly zero-energy building (NZEB) in Bulgaria, and consequently, in other countries in Eastern and Southern Europe.

Chapter 12 "The public in search of identity: new symbolism in urban spaces. A study of central squares of Balkan capitals" is devoted to thorough analysis of the process of building new political and cultural identities in Western Balkan. This chapter presents a study of identity formation, specifically how identity in four Western Balkan countries is expressed in significant place. The alterations of central squares of capitals, regarded as tangible expressions of national identity, have been studied in four small countries; Albania, Kosovo, Macedonia, and Montenegro. The study focuses on how selective remembrance and active forgetting played a role in the development of main squares, how ambiguous spaces have been created with also banal practices, and how the squares can be seen in view of globalised identities and cultural hybridities. Cases of cities of Tirana, Pristina, Podgorica, and Skopje illustrate this process.

Last chapter "Professional associations as public actors in the formulation and implementation of spatial development policies. Key study monitoring and evaluation standard for the urban environment of Sofia" deals with the gradual shifts of the role of planning profession, its institutions and planning in general in Bulgaria. It introduces the conceptual context of the transactive planning in spatial development and lower level institutional thickness. Thorough analysis of the key studies of The Association of Bulgarian Urban and Regional Planners (ABURP) demonstration projects "Share the Neighbourhood" and "Set an Urban Standard" in Sofia are the main scopes of this chapter.

Chapter 2
Innovations and Changing Role of Public Sector in Spatial Development Strategies: Problems and Challenges for Local and Regional Development in Central and Eastern European Countries After 2020

Milan Husár, Maroš Finka, Ľubomír Jamečný, and Vladimír Ondrejička

2.1 Introduction

The global transformation processes of the society at the edge of twentieth and twenty-first century accelerated by the transition processes in majority of the countries in Central and Eastern Europe represent huge challenge for spatial development. Moreover, they deeply concern not only the development of the physical and functional structures, functioning and qualitative and quantitative changes in the Central European cities and regions, but they represent crucial changes in the nature of the spatial development management as well. The shift from government to governance framed by the movement towards civil society development is combined with the fuzzification and softening of the borders between functional and administrative spatial units, with the multiplication of the relevant actors in spatial development and increased spatial effects of their individual decisions, with the regionalisation of local policies, as well as state policies and with the shift from ecologisation towards economisation of spatial planning (Finka et al. 2015). Spatial development is about intensive interplay between all hierarchical levels, between different temporal dimensions, between public and private actors. The public sector representatives are more and more diverse, going far behind the governmental and self-governmental bodies. They play specific role in the management of spatial development being more and more limited in their operational space by strong economic interests, interests expressed by broad public, and responsibilities for safeguarding societal values and sustainability. The role of public authorities is changing

M. Husár (✉) · M. Finka · Ľ. Jamečný · V. Ondrejička
Spectra Centre of Excellence of the EU, Slovak University of Technology in Bratislava, Bratislava, Slovakia
e-mail: milan.husar@stuba.sk; maros.finka@stuba.sk; lubomir.jamecny@stuba.sk; vladimir.ondrejicka@stuba.sk

© Springer International Publishing AG, part of Springer Nature 2019

M. Finka et al. (eds.), *The Role of Public Sector in Local Economic and Territorial Development*, EAI/Springer Innovations in Communication and Computing, https://doi.org/10.1007/978-3-319-93575-1_2

in the development and implementation of spatial development strategies as their position among public sector representatives is changing (Špaček 2015). The examples from Slovakia and other European Countries show the problems and challenges connected with this development as well as the whole range of creative solutions and possibilities to strengthen the quality and efficiency of public interventions towards meeting the goals of spatial development.

2.2 Societal and Economic Transformation Processes Framing the Innovations and Changing the Role of Public Sector in Spatial Development Strategies

Spatial development strategies, the processes of their designing, implementation, and monitoring are experiencing a variety of challenges. The following chapter is attempting to frame this transformation process by discussing the main characteristics of the transformation. Economy in the EU member states is affected by wide-reaching changes of the production patterns, closely linked to the development of knowledge-based economy characterised by greater dependence on information, knowledge, and high skills and increasing need for ready access to all of these by the business and public sectors (OECD 2005). These changes create a number of implications for the planning systems including change in the value systems, need for open data, and innovations in governance mechanisms or perception of administrative borders (Ondrejička and Finka 2016).

Reflecting the rising awareness about the limitations of resources for societal development and about the role of humans in global climate change, environmental values are coming increasingly to the foreground. The need to balance economic, monetary-based valuation with environmental values is mirrored in the development and broad implementation of the concept of ecosystem services as a way to express, calculate, and to assess the value of the benefits resulting for humans from the presence and active use of natural ecosystems as the part of social ecosystems of the cities and regions.

Typical feature of recently dominating post-industrial economy is its orientation towards products with higher added value. This signifies that the companies are trying to give the product the most enhanced value before its selling in two directions—maximising the proportion of non-material part of the product value resulting from the human work and knowledge and, on the other hand, maximising value by meeting individual demands of the potential buyers. The shift of individual value systems towards the environmental values derives the pressure towards more environmentally friendly products and production processes. The use of closed, environmentally friendly technologies across all production sectors and human activities (e.g. transport, services, energy supply) allows increased polyfunctionality of the spaces and spatial proximity of different functions creating the basis for the implementation of the compact city concepts, new urbanity, economy of spatial proximity, and others.

This implies significant changes in spatial organisation of the production, material flows, and ties in the production sectors and among them and other functions in the cities and regions. The twentieth century experienced the greatest technological advancements in human history, and this rapid advancement continues at the turn of centuries expanding rapidly into the non-production sectors and people's lives, be it in the form of personal computers, communication technologies and technological gadgets, smart home technologies or advanced technologies using renewable energy in the households. The most important shift at the turn of centuries is the change in the dominancy among the production sectors. The leading position of the industry has been replaced by the services and fourth sector representing the production of innovations of the highest level in the science, research, and development. The triple helix universities-industry-government became significant concept applied in the territorial organisation of the production (Zajko and Brighton 2015). The new organisational patterns of production sectors—the production clusters determined by the development and spread of innovation more than simple horizontal production cooperation ties became more and more independent from traditional territorial organisational patterns expressed by the administrative borders.

With the forthcoming automatisation and robotisation, the character of work of the dominant part of population is changing. The intellectual work dominates above the manual, the place of work is more and more independent from the place of goods' production, even provision of the services. This requires changes in the quality of working places, their localisation more and more independent from the organisational patterns of the demands or location of the natural resources. On the other hand, the quality of environment for the labours, quality of their living preconditions influencing their productivity became the most important localisation factor.

This shift is accompanied by significant changes in the consumption patterns related to the changing behaviour and demands of citizens as consumers. The consumer demands are not decreasing, but are changing for instance preferring more eco-friendly, green products, but more individual (original, to the specific needs of consumer neaten) products as well. Consumers are increasingly more likely to prefer renewable materials of higher quality and more eco-friendly recyclable products. These amplified demands are reflected also in people's demands on space and their living and working environments. With increased work productivity and overall economic status, citizens can choose the place to live and work and are less likely to look on the price, but their preferences are increasingly focused on high quality of public spaces, social services, and transportation options. As a result of shifts in production and consumption patterns, there are ongoing changes in the societal organisation. Working patterns are transforming, the work in the production became more individual, fewer citizens working regular nine-to-five jobs, more people are working from home, there is an increase in number of digital nomads, i.e. people able to work from anyplace with stable internet access. At the same time, the productivity of work is increasing and together with trends in automatisation people are working less hours and are spending more time at home with their families realising their family and recreation activities. For planning it means change of peak hours for transportation, but also increased demands of the space of people requiring high quality services and safe and liveable public spaces. These shifts are demonstrated also in energy use patterns and peak hours as a result of different time spent at work rendering the traditional morning and evening peaks in energy consumption inaccurate. The development of the information and communication technologies brought the expansion of the existential space of people into the virtual space. More and more human activities take place in this pace starting with social networking, via shopping and ending with many public services (Jaššo and Petríková 2016). The open accessibility of virtual spaces together with growing mobility of the people lower the feeling of territorial belonging. The free choice of working place, place for dwelling, leisure activities in the combination with growing importance of specific individual demands and better access to their fulfilment due improving of economic capacity of the citizens strengthens extension of individual existential spaces of the people beyond traditional borders of territorial units (municipalities, regions, national states). This, together with similar processes in the production sectors leads to the fact that the current space organisation is more and more determined by the fuzzification of the territorial organisational units.

Strengthening the civil society is another characteristic of twenty-first century framing the transformation processes for changes of the role of public sector in spatial development strategies. Significant is the shift from representative towards direct democracy with involvement of broad scale of actors into the decision-making processes, growing engagement of citizens in the numerous NGOs and their involvement not only in the decision-making processes but in the realisation of projects implementing public development strategies too, with the engagement of private sector as the key player of spatial development into the development and implementation of public policies (Schoeffel et al. 2014; Ondrejička and Ondrejičková 2014). People are more empowered as a result of better participation and governance processes and due to the self-realisation of their power to affect the decision-making. The hierarchic territorially organised system of governing public bodies as the leading actors in the decision-making is going to be transferred into the system of polycentric multilevel governance based on multi-actors decision-making processes.

The development of polycentric multilevel governance systems together with the extension of existential spaces of the citizens beyond of administrative borders of their municipalities, regions, and national states and fuzzification of the borders of functional spaces leads to lowering the justness of the governance patterns based on territorial responsibility and public participation based on territorial belonging (Finka & Kluvánková 2015).

The processes of environmental impact assessment and strategic impact assessment and the Aarhus Convention have created fertile ground for civil engagement and past years are clearly showing the benefits from such public involvement (Jaššo et al. 2014). No plan or strategy is accepted within particular level of participation, but there are still deficiencies especially in governance modes which are not fully developed in many countries, yet and there is still a lot of opportunity to make more room and create conditions for better reflecting the civil society in the decision-making. The EU directive on strategic environmental assessment had opened the space for participation for all people independently from their citizenship and it is in the same time example of the reflection of borders' fuzzification up to globalisation of the environmental effects.

Institutional settings are being steered from government towards governance modes changing the role of the state and eliminating its power monopoly. The power is being shifted vertically to international and sub-national levels with regional and local levels gaining more jurisdictions, and horizontally focusing more on cooperation mechanisms and agreements among branches of public policy. This shift is supposed to make the governing mechanisms more effective and efficient by more inclusionary nature and giving some power traditionally comprised of the national level on regional and local levels (Jaššo 2008).

2.3 Challenges for Economic and Territorial Development Related to the Role of the Public Sector

The challenges for the economic and territorial development in Central Europe can be grouped into two main categories:

- Substantial, related to the development as the object of the policy
- Processual, related to the decision-making processes and governance

The former factors signify concrete issues dealing with the development as an objective, and it includes practical challenges traditionally dominating in the debates on optimal development, while the latter group reflecting the challenges in governance mechanisms and the decision-making processes (Ondrejička and Finka 2016) became to be the focus of professional debates at the turn of centuries (Table 2.1).

Main Substantial Challenges for Local and Regional Development Strategies in Central Europe

Development of knowledge-based society is a precondition of economic and social development in the EU. As stated above, the transition towards knowledge-based society is affecting both, private and public sector, and majority of their fields. New spatial organisational patterns of the production sectors, expansion of the existential space of the people, fuzzification of the borders of territorial units, qualitative and quantitative changes in the composition of the functions in the cities and regions, and shift of many traditional urban functions into the virtual space, new demands on quality of the environment including public spaces are only some of the implications of the current knowledge-based society development directly addressing spatial development strategies.

Demographic and social changes are viewed often as threats to the European continent, but they also pose a huge opportunity. Less children are being born, population is ageing, and the average person's age is increasing. This brings not only the issue of labour force shortages, but also it changes the demand for social services. Schools and universities are threatened by decreasing number of students and

Table 2.1 Challenges for economic and territorial development related to the role of the public sector (elaborated by the authors)

Main substantive challenges for local and regional development strategies in Central Europe related to the development as the object of the policy	Main processual challenges for local and regional development strategies in Central Europe in the field of decision-making and governance
Development of knowledge-based society as the precondition of economic and social development in the SR	Need to reflect new policies and approaches at international level
	Need to transfer macro-economic achievements into the quality of life of the people
Demographic and social changes	Need to face the lack of public resources and to look for their more efficient use and combination of resources
	Need to move from passive to proactive social policies
Migration of young people from the regions, interregional and international migration of high qualified labours versus increasing the mobility of labours	Change in the understanding of the content of local development policies towards economisation of them
	Need for flexibility of the strategy in reaction on identified needs, reflexive assessment of the needs and problems, efficiency
Forthcoming globalisation, integration, and openness bringing growing exposition to external stress	De-bureaucratisation and implementation of managerial concepts in regional development management
	Shift of the regional development policies from the orientation to the lowering the disparities towards capitalisation of specific potentials and support of the engines of the development
Growing complexity and vulnerability of socio-ecosystems, specific social groups, economic sectors, and landscape types	Development of new system frameworks for efficient use of new instruments of the regional development policy and their permanent innovation
	Change of the content and de-formalisation of the planning documents and increase of their efficiency and implementation of new innovative instruments in the RD policies
Growing fuzziness of territories and processes, clustering across physical spaces	Consequent implementation of the principles of subsidiarity in the defining regional development strategies, their goals, and implementation tools including financial tools
	Cross-sectoral coordination and subordination to the integrative policies (the same on EU level)
Growing role of individual decisions in the combination with growing number of actors in decision-making processes	Interaction and partnership with regional development actor in both phases—definition and implementation of the development strategies
	Affordability of the resources for RD for key target stakeholders

(continued)

Table 2.1 (continued)

Main substantive challenges for local and regional development strategies in Central Europe related to the development as the object of the policy	Main processual challenges for local and regional development strategies in Central Europe in the field of decision-making and governance
Growing unpredictability of development processes	Multi-actors decision-making/multilevel polycentric governance
	Feedback from regions and flexible reactions to identify problems, failures, and success
Growing spatial disparities in the context of peripheralisation and centralisation and disparities in the communities	Availability of needed information for efficient planning, programming, management, and monitoring of real development processes across all hierarchic levels
	Increase of efficiency and electronisation of governance services
Losing regional and local identity (as an effect of growing fuzziness, mobility, extension of existential spaces)	Correct definition of strategy implementation indicators and transparency of effect assessment
	Synergy effects as the motivation to closer horizontal and vertical cooperation, addressing the benefits to the resources of savings
Growing fuzziness in territorial belonging bringing losses of territorial legibility in the right on participation on decision-making	Conceptual shift from integrated to integrative policies including social inclusion of disadvantaged social groups
	Development policy focused on capitalisation of potentials as the tool to face the problems/evidence-based planning
Multidimensional definition of functional spaces versus administrative definition of territorial units	Use of potential of cooperative structures and capacities in the regions (more than 90% of municipalities in informal cooperative structures)
	Use of potential of engagement of regional development stakeholders supported by proper communication
Big dynamic in economic development—potential to invest into local development in the phase of dynamic economic growth and use of European Structural and Investment Funds	Open fuzzy system of spatial management (across administrative borders, across timelines)
	Planning for fuzzy spaces
Effect of the short-term effects-oriented decision in market economy	Implementation of community-based approaches in the city management—from government to governance
	Self-control processes in socio-ecosystems and in the same time safeguarding the sustainability

(continued)

Table 2.1 (continued)

Main substantive challenges for local and regional development strategies in Central Europe related to the development as the object of the policy	Main processual challenges for local and regional development strategies in Central Europe in the field of decision-making and governance
The uneven distribution of the benefits and lost from the economic development (loss of working places, public goods…)	Growing role of self-learning processes in the governance implemented by bottom-up as well as top-down processes
	Overlapping of different planning cultures
	Softening of planning approaches/from the definition of target quality to framing the processes
	Softening of planning tools/from legally binding to voluntary agreements—from hierarchy to partnership

the pressure on number and capacity of social workers and social facilities is higher. This significantly changes the basic axioms in urbanism and spatial planning regarding the norms for optimal capacities of social facilities in the settlements. In the job market, there are already calls for increasing the labour force access for workers from non-EU countries due to gradually more acute scarcity of workers in many production and service sectors. In Slovakia, for example, the economy is highly built on automotive sector, lately with the construction of Jaguar Land Rover factory and labour market already cannot provide enough skilled labour force. These factories and the related subcontracting and supplying companies require labours skilled in working with automatic machines and the country has a hard time providing these. This demand derives interregional and large-scale migration connected with the demand on proper transport, dwelling and social infrastructure, in some cases very specific respecting the needs of immigrants with different cultural background. With the increased mobility of the capital invested in industrial production, the question of the efficiency of the investment into such infrastructure appears as the working places can be very temporal. Something similar is taking place in many other production and service sectors in various European countries.

As a result of the situation at the labour market, young people and skilled workers motivated by the desire for better life increased their flexibility, used to migrate out of their home municipalities within the regions, national states, and internationally. This increased mobility of labour has an impact on real estate market—change in preferences regarding the flats ownership mode—from use of own flat or house towards renting them. On the other hand, there is increased demand on higher quality of social facilities, transportation, recreation opportunities, and pressures for more open labour markets and decreasing the labour entry barriers. The migration patterns motivated by the accessibility of the working places, services, and leisure time activities are changing due to improved transportation (quicker and cheaper) and due to permeable administrative borders (e.g. made possible in the Schengen space). All these challenges the usual notions of transportation and migration patterns and spatial planning needs to pro-actively reflect these patterns.

The accompanying phenomenon of forthcoming globalisation, integration, and openness is also growing exposition to external stress. Cities, as social ecosystems in their complexity and openness to the interaction with the surrounding ecosystems and social systems, are exposed to various types of natural and man-made stresses (Finka et al. 2015). With cities' growth in size and density and the increasing interconnections between systems within the city and within the regions, their complexity and entropy are growing and so does its vulnerability which is accelerated by their development. Cities cannot be fully rationally regulated, but they have to be perceived as living laboratories and their self-learning capacity has to be reflected in the planning across all scales.

Planning is turning into more negotiation and mediation of various interests in the space. The role in individual decisions in the combination with growing number of actors in decision-making processes is increasing. Thus, planning is no longer about directing the development, but gradually about fostering consensus building and looking for ways to harmonise the interests of stakeholders in sustainable way while following the principle of democracy (Petríková and Ladzianska 2012). The main challenge is no longer about designing blueprint plans of how the space will look like in broader time horizon, but about facilitating the dialogues among the representatives of the public sector, private sector, and NGOs by including them in the governance modes and enabling their communication and negotiations. Planning systems have to embrace this role and to be assertive enough to claim their new significant role by staying accountable and responsive.

The increasing complexity, the multiactors' decision-making together with growing exposure to external shocks are resulting into the growing unpredictability of development processes of the cities and regions. Dynamics and unpredictability are basic characteristics of urban and regional socio-ecological systems, and they are manifested in high scale of fluctuations due to its emergence characteristics. The public sector with its dominant role in the management of the spatial development processes needs to start reflecting this fact more in its approach, methods, and tools used in the management and planning of the development processes (Dostál et al. 2010). It has to acknowledge that the complex systems are less and less predictable as the role of individual development actors with their unpredictable behaviour grows. It means the detailed descriptions of willing future development in the form of blue prints have to be replaced by the definition of limits for different development trajectories (influenced by dynamically changing framework preconditions), safeguarding the achievement the development goals including sustainability, and by the definition of the measures to avoid the development trajectory overstepping the borders of the by the limits defined development corridor.

In spite of huge public intervention at the EU as well as at the level of EU member states spatial disparities have increased in the CEE countries in the past years (Lang 2015). It results in growing territorial as well as social polarisation. This trend, observed among European countries and among their regions, shows the problems of efficiency of spatial development policies across all levels. Territorial cohesion had been on the EU agenda for more than a decade, and these differences are taking place still. Reflecting them in the development policies has to become one of the top priorities for governments and the decision-makers have to focus on delivering effective results to stay competitive (Špaček 2015).

One of the undesired effects of development policies in the past had been losing the regional and local identity. Regions and authorities are often focusing excessively on delivering positive economic results which bring about foreign investments changing the economic and social structure of the regions, often changing their identity and distinctiveness. On the other hand, the loss of identity takes place as an effect of growing fuzziness, mobility, extension of existential spaces (Poklembova et al. 2013).

Growing fuzziness in territorial belonging connected with losses of territorial legibility in the right on participation on decision-making represents one of crucial challenges for spatial planning especially its participative concepts. In addition, with increased border permeability the functional spaces of individuals are expanding and people inhabiting territorially on a daily basis several administrative jurisdictions and in many of these they do not have access to the decision-making processes. As a result of this is inability to be part in any sort of decision-making (e.g. voting rights, participation in activities) in the places they are belonging to with their dominant life activities and are excluded as illegitimate participants. There is a contention between the multidimensional definition of functional spaces versus administrative definition of territorial units. The administrative definition is rigid and does not reflect the new spatial reality and freedom of choice of people's functioning. This need for change needs to be reflected on legal scale and requires a change of the basic idea of territorial administrative units and forms of belonging (Husár et al. 2017).

Since the global economic crises, many European economies have relaunched the dynamics of their economies, and the EU is enabling this growth by making it available to invest into local development in the phase of dynamic economic growth and use of European Structural and Investment Funds. This form of economic assistance offers considerable boost and opportunity for public spending for public sector and cooperation with the private sector on public projects. Never in the history was it easier for public authorities to apply for financial help and develop the public projects. At the same time, though, this brings the need to recognise this opportunity and use the financial aid to support the local and regional growth not only by fighting negative regional disparities but much more capitalising their territorial potentials as the tool to face the problems as means to evidence-based planning. This helps making plans based on reflecting the actual situations and using the available territorial and social capital (the so-called smart specialisation strategies using the language of the EU) (Husár 2015).

Meanwhile there is a need to think ahead and sustainably about the regional growth and consider the long-term effects of decisions in the market economy. The threat of effects of the short-term effects-oriented decision in the market economy had been manifested at the dawn of the global economic crisis which was a result of the moral hazard of big economic players, but also partially public authorities neglecting this crucial point during their projects. Responsibility and accountability have to become deeply rooted in the ethical codes of conduct for public authorities, and rigorous monitoring of the foreseen long-term effects has to be closely examined in all public projects. There is a need to move from formal indicators to rational and essential indicators of project sustainability and the effects on the quality of life.

Lastly, the threat uneven distribution of the benefits and lost from the economic development such as the loss of working places or public goods needs to be kept in mind during the creation of development plans. Economic indicators and benefits have to be complementary to the social and environmental dimensions of public projects in order to release the synergies and deliver good sustainable and beneficial development for citizens. During economic restructuring, there has to be transitory period to avoid economic and social shocks, for example, in job losses to avoid threatening the livelihoods of people. These had been stumbling blocks of the economic transitions in the 1990s in the CEE countries during which thousands of people lost their jobs and future prospects due to indifferent interventions of public authorities.

Main Challenges for Local and Regional Development Strategies in Central Europe in the Field of Governance and Decision-Making

This section is describing the key challenges for local and regional development strategies related to the processes of governance, especially of decision-making framing the process of strategic planning. Planning is never static and general and situated in a bubble, it is highly dynamic process, and it is always place specific. The plans and strategies need to be responsive to the local needs, to the needs of the local and regional conditions, the specifics of the local and regional economy, and also consider the wider territorial and economic relations (Jamcčný and Husár 2016).

In globalised world, these relations require the need to reflect new policies and approaches at international level, in case of the CEE countries the EU institutional settings and strategic documents and other, such as UN resolutions, Millennium Development Goals, etc. Cities as complex social ecosystems have emergent properties, and all settlements are positioned in territorial systems which are mutually affecting each other. Higher hierarchical levels have impacts on lower levels and vice versa. Therefore, taking into account in the planning processes of local and regional units these higher levels and set them into specific local contexts. In the similar vein, planners and other stakeholders involved in the planning processes have to transfer macro-economic achievements into the quality of life of the people.

In the face of resource scarcity (financial, human, material resources), the public sector needs to face the lack of public resources and to look for their more efficient use and combination of resources. With the highest rate of urbanisation in the history in terms of impacts on the resources, this threat becomes very real and poses an utmost challenge when designing sustainable works. Resources are limited while many of them are not renewable and relying on technological progress to replace them in the future is irresponsible conduct. This has to be accompanied by the rising awareness of the public about the limitation of the resources of all kind and by changing the behaviour of the people towards more sustainable and efficient use of all resources starting with the non-renewable and renewable natural resources up to the human resources. In this the role of public sector is absolutely dominant.

In the realm of social policies, we need to move from passive to proactive social policies. Above we have mentioned the need to react properly on demographic trends of population ageing, or long-term unemployment as the effect of economic transformation, but social policies need to go beyond. The aim should be not only to reflect this trend in the strategies, it should be first of all to stop or, at least to lower the negative trends and secondly to focus on lowering or mitigating their negative effects.

The philosophy in dealing with the problems should be changed from focusing on the effects towards focusing on the background of the problems, e.g. besides specific planning for elderly population or social groups with special needs the planning processes need to be inclusive and consider all members of the society including the disabled and economically challenged groups of citizens and enforce solidarity measures in order to foster social capital of the place (Finka et al. 2012).

There has to be a shift in the understanding of the content of local development policies towards economisation of them. There is a load of critique concerning the concept of ecosystem services as they are trying to monetise the nature, but in the neoliberal economy there had to be developed methods to allow the comparison and balancing different values and effects and in the end of the day to capitalise ecosystem services as important part of the local and regional economic performance.

The planning procedures and processes need to be flexible in reaction to the identified needs. As the framework conditions for the development can change and are difficult to predict the planning processes have to be designed to be ready for flexible reflection of the changing development dynamics (Finka et al. 2015). The assessment of the needs and problems should be reflexive and planning documents have to be devised so that they can properly accommodate the change in response to the results of the monitoring and re-assessment.

One of the repeating pleas to the development planning is excessive bureaucratisation resulting oftentimes in delays and lengthening of the development processes. On a more general note, this can be considered as one of the main characteristics of the public sector. Hence one of the prerequisites of effective development of strategies is de-bureaucratisation and implementation of managerial concepts in local and regional development management. This issue had been demonstrated by David Cameron in his statement from September 6th, 2012 when from the position of Great Britain's Prime Minister he said that bureaucracy is holding economy back and dealing with it we need "getting the planners off our backs" (The Guardian 2012). Although the statement had launched heated defensive discussions among planners, planners and decision-makers need to consider the partial truth in the quote. Perhaps a more appropriate response from the public sector should have been deeper reflection and reconsidering the necessity of complex and often redundant bureaucratic processes.

The basic idea of the EU cohesion policy in the previous periods had been to support the less developed regions only and not supporting the better-off ones. Nevertheless, these regions need to start to be considered as true engines of the development and be supported as well. The regional development needs to be shifted from the orientation to the lowering negative disparities towards capitalisation of specific potentials and support the drivers of the economic development to release the synergies among regions and finding the proper balance between cooperation and competition.

Another crucial challenge is de-formalisation and changing the content of the planning documents in order to increase their efficiency. This can be done by implementation of new innovative instruments in the local and regional development policies. Regarding the planning processes, consequent implementation of the principles of subsidiarity in the defining regional development strategies, their goals, and implementation tools including financial tools have to be introduced and enforced in the planning practice.

In the quest for effective planning, the planning needs to recognise the need for cross-sectoral coordination and subordination to the integrative policies. Planning challenges are not existing in a vacuum, but majority of them are highly interlinked and are affecting each other. Same goes with the planning solutions which are influencing other areas of public policies. Nearly no issue stands in its own and plans and projects need to be horizontally coordinated. This requires, among others, fostering continuous dialogue among branches of the public sector and their cooperation to break the institutional silos which is also linked to the issue of de-bureaucratisation and breaking the barriers among public departments and institutions.

One of the basic characteristics of the current planning and decision-making processes is the increased number of actors within the regional development (Ondrejička and Finka 2016). Formally, this issue had been tackled by number of directives on public participation in the previous years, but what is missing is the interaction and partnership with regional development actors in both phases—definition and implementation of the development strategies. In practice, the actors are formally included mainly in the definition of development strategies and their involvement their implementation is seldom. The shift has to take place in thinking about these processes and recognising the significance of incorporating the multitude of actors on the whole course of regional development plans designing and implementing.

This is connected to the topic of affordability of resources for key target stakeholders which remains an issue in many EU regions where the public authorities and other public stakeholders have limited resources to participate and fulfil the requirements (for instance, the co-financing) which as a result discourages their involvements in the projects and decision-making processes.

One of the tools to foster more inclusive and effective planning is by enabling the multi-actor decision-making via implementing the concept of multilevel polycentric governance (Finka et al. 2015). This concept is based on vertical and horizontal coordination of actors in the decision-making and creating arrangements for involvement of large number of stakeholders as partners in the dialogues on regional development.

The concept of multilevel governance is trying to capture the complexity of horizontal relations among multiple levels of governance within the EU and in the polycentricity adds the dimension of cross-sectoral coordination based on research on natural resource management and the interlinks between a variety of factors influencing the success of the management.

Multilevel polycentric governance also helps to obtain and include the feedback from regions and enables flexible reactions to identify problems or failures as it has adaptive capacity allowed in its flexible structures (Finka et al. 2013). Feedback is obtained from monitoring processes and provides information on the plan or strategy implementation and its effectiveness, and these are used as inputs into strategy modification or into the process of new strategy designing.

Efficient planning is highly dependent on the quality and accuracy of the input information which must be taken into account during planning, programming, management, and monitoring of real development processes across all hierarchic levels. This information has to be available and up-to-date and, moreover, the planning processes have to be designed so that information can be effectively utilised. The data availability is fostered by open data policies and public sector must make this data accessible to planners and other actors involved in planning processes.

Electronisation of the governance is an ongoing trend and national governments of the EU member countries. It had been investing large amounts of money into the projects of electronisation. Some countries such as Estonia used electronisation as the key idea during their presidency of the Council of the European Union in the second half of 2017, where they demonstrated their success story. Not all countries are doing so well, though, and therefore the objective of increasing the efficiency and electronisation of governance services has to be on the top of agenda. Effective electronisation has to enable easier access to the public sector request and involvement in the governance processes and foster availability and openness of data.

The importance of monitoring processes and mechanisms was mentioned before, but a prerequisite for effective monitoring is the correct definition of strategy implementation indicators and transparency of effect assessment. Indicators need to be both quantifiable and, more importantly, qualitative in order to objectively assess the strategy progress and to evaluate the impacts on the quality of life of citizens. Strategies need to shift from numbers-only indicators such as number of positively affected citizens, to trying to evaluate the overall impacts. These are difficult to define and even more difficult to calculate, but the strategy success or failure depends on them.

Synergy effects need to be perceived as the motivation to closer horizontal and vertical cooperation, addressing the benefits to the resources of savings. Savings in the public sector had become key objective, but it is often misleadingly recognised in monetary terms only. From the point of view of the resource scarcity, the human and natural resources are more important for their limited availability and enabling synergic effects from mutual operation of horizontal and vertical levels.

Policies should make a conceptual shift integrated to integrative policies including social inclusion of disadvantaged social groups. Integrative policies go beyond the usual definition of social policies aimed at elderly people to genuine efforts to be inclusive for all disadvantaged social groups which must be perceived as a source of immense social capital.

Regional development strategies have to maximise the use of potential of cooperative structures and capacities in the regions (e.g. in Slovakia more than 90% of municipalities are part of informal cooperation strategies). The regions have to find the right balance between cooperation and competition in order to use the potentials of theirs and of their neighbouring regions as a competitive advantage in the global network of regions racing for investments as means to grow. Such cooperation, though, has to shift from mostly formal and purposefully built for applying for funding, to true cooperation where all partners are contributing and reaping the rewards for their mutual efforts, even if it means the actual decrease in the number of cooperative structures.

As already mentioned above, the increasing permeability of the administrative borders driven by growing mobility of persons and goods contributes to the fuzzification of border and administrative units. The fuzzy borders require open fuzzy systems of spatial management across borders and timelines as the planning cannot be easily contained anymore into neatly delineated containers as stand-alone units where activities take place. Planning for fuzzy spaces need to account for overlaps of administrative units and among branches of public policy. These overlaps are spaces where the current issues are being resolved, and it is here where the potential answers for current challenges are laying. Complex problems such as migration or climate change are affecting vast spaces which formally stretch across many formal territorial units and the effective responses to them have to cover them fully together with legal jurisdictions to have power to create appropriate resolutions.

Community-based approaches to the implementation in the spatial development management need more promotion in the planning practice in the form of shift from government (top-down, hierarchical approaches to directive rule) to governance (more open, cooperation, coordination, and inclusion-based approaches to spatial management). This is closely linked to the self-learning capacity and processes in the governance implemented by bottom-up processes combined with the existing top-down processes (Schoeffel et al. 2014). The objective is to find harmony between these two approaches and design specific, context-based plans and strategies. Cities and regions as social ecosystems also have capacity to self-control the development processes and at the same time to safeguard the sustainability of their functioning. These attributes require more open and more flexible approaches and ways of governing to be fully utilised and allowed to operate.

The planning system vary in each country, however, based on the overlaps of the characteristics of the planning systems these can be grouped into planning cultures. Planning cultures have numerous overlaps and can mutually contribute to each other, for instance elements of ex-post planning are becoming increasingly more relevant in order to incorporate the rapidly changing conditions of regional development plans and assessing projects against the existing plans and regulations (Jaššo et al. 2014). Therefore, planning cultures and planning systems require remaining being open and use the strengths of self-learning and adjusting.

Lastly, the last trend we are listing is softening of the planning approaches and the planning tools. Planning approaches need to be designed holistically from the definition of target quality to framing the processes to achieve it. Within all phases of the planning process, there needs to be ensured appropriate methods to include the multi-actor decision-making and allow for incorporating well-articulated needs and information. These characteristics have a potential to greatly help to design and implement effective and efficient plans and strategies in sustainable way. Softening of planning tools requires a change from legally binding (laws, regulations, strictly enforced compulsory plans) to voluntary agreements and a shift from hierarchy to partnership. In partnerships, all partners are working together following a shared vision and are motivated to succeed and commonly reap the benefits of good planning and implementation of plans and strategies.

2.4 Conclusions

In spite of many differences from country to country, the processes of the EU integration and globalisation frame the set of principal problems and challenges for local and regional development in Central and Eastern European countries. They are connected with overall transition process of the EU countries towards civil society, knowledge-based economy and society, circular low carbon economy, with the common challenges for them like limited resources and sustainable efficient use of them, climate change adaptation, informatisation, electronisation, robotisation, and others. At the same time, there are specific challenges for them linked to common high complexity of transformation processes and dynamics of them typical for Central and Eastern European countries. Both the common and the specific problems and challenges are linked to the spatial development as the object of the managerial interventions of the public sector as well as the system of spatial development management itself and the role of public sector in it.

Identified deep changes and principal challenges need proper responds in the content and in the processes of the elaboration of spatial development strategies (as integrative and integrated strategies), as well as particular strategies for the development of spatially relevant fields of societal life. Conceptual innovations have to be followed by the substantial, instrumental, and processual innovations reaching from institutional, via behavioural up to technological. They should be focused on higher efficiency and justness of the spatial development management not only in the sense of social justice but in the sense of appropriation to new needs and new preconditions framing spatial development. The strategic planning, playing the key role in spatial development management in decision-making sphere for the development and spread of innovation across all parts of spatial development management including the scoping, diagnosis, planning, designing, implementation, and monitoring as will be shown in the other chapters of this book.

References

Dostál I, Jedlička J, Kortanová J (2010) Planning for sustainable development of towns and villages. In: Adamec V, Jandová V (eds) IV Czech-Slovak scientific conference "Transport, Health and Environment". Transport Research Centre, Brno

Finka M, Kluvánková T (2015) Managing complexity of urban systems: a polycentric approach. Land Use Policy 42:602–608

Finka M et al (2012) Bezpečnosť ako kvalita priestoru—úvod do problematiky [Safety as spatial quality—introduction into the topic]. ROAD/Centrum urbánnej bezpečnosti, Bratislava

Finka M et al (2013) Governing under the complexity and uncertainty. In: Kluvankova-Oravska T, Jilkova J (eds) From governing to governance resonsidered. Verbum, Ružomberok, p 2013

Finka M, Kluvánková T, Ondrejička V (2015) Concept of polycentric governance for fuzzy soft spaces as a challenge for Central European Peripheral Spaces. In: Lang T, Henn S, Sgibnev W, Ehrlich K (eds) Understanding geographies of polarization and peripheralization: perspectives from Central and Eastern Europe and Beyond. Palgrave Macmillan, New York, pp 309–322

Husár M (2015) Cities of tomorrow? Smart city as an imperative for sustainability in the future. Bratislava: In Terra Spectra STU. Planning Studies., 2015. Zv. 7, 1

Husár M, Varis SC, Ondrejicka V (2017) Analysis of the planning education in the light of the contemporary trends in planning. IOP Conf Ser Earth Environ Sci 95(5):052005. https://doi.org/10.1088/1755-1315/95/5/052005

Jamečný L, Husár M (2016) From planning to smart management of historic industrial brownfield. Procedia Engineering. Prague: World Multidisciplinary Civil Engineering-Architecture-Urban Planning Symposium 2016, WMCAUS 2016

Jaššo M (2008) Cross-border cooperation challenges: positioning the Vienna-Bratislava region. In: Cross-border governance and sustainable spatial development: mind the gaps! Springer, Berlin

Jaššo M, Petríková D (2016) Place attachment and social communities in the concept of smart cities. In: Leon-Garcia A et al (eds) Smart City 360°. SmartCity 360 2016, SmartCity 360 2015. Lecture Notes of the Institute for Computer Sciences, Social Informatics and Telecommunications Engineering. Springer, Cham, pp 721–728

Jaššo M et al (2014) Plánovací systém priestorového rozvoja a plánovacia kultúra v slovenskej republike [Planning system of spatial development and planning culture in the Slovak Republic]. In: Schoeffel J, Jamečný L, Ondrejičková S (eds) Participatívne plánovanie na úrovni samospráv [Participative planning on municipality level]. SPECTRA Centre of Excellence, Bratislava

Lang T (2015) Socio-economic and political responses to regional polarisation and socio-spatial peripheralisation in Central and Eastern Europe: a research agenda. Hungarian Geographical Bulletin 64(3):171–185

OECD (2005) "The Measurement of Scientific and Technological Activities: Guidelines for Collecting and Interpreting Innovation Data: Oslo Manual, Third Edition" prepared by the Working Party of National Experts on Scientific and Technology Indicators, OECD, Paris, para. 71

Ondrejička V, Finka M (2016) Innovative approaches to strategic socioeconomic planning of territorial subjects. Vysoka Skola Ekonomicka v Praze, Praha, pp 176–188

Ondrejička V, Ondrejičková S (2014) "Trenčín si Ty" [You are Trencin]—best practice of participatory planning in Slovak Republic. In: Finka M, Schoeffel J, Ondrejička V (eds) Participative planning in planning culture of Slovak Republic and Switzerland. IRAP, Raperswill, pp 41–61

Petríková D, Ladzianska Z (2012) Social and cultural issues and participation in brownfield redevelopment. In: Petríková D, Vojvodíková B (eds) Brownfields—handbook BROWNTRANS. VŠB—Technical University of Ostrava, Ostrava

Poklembova V, Maco M, Ondrejicka V (2013) Ako spravovať priestor, o ktorý sa delíme? Teória commons a verejné priestory [How to manage the space we are sharing? The theory of Commons and public spaces]. URBION, Bratislava, pp 28–31

Schoeffel J et al (2014) Participative planning in planning culture of Slovak republic and Switzerland. 1.vyd. Rapperswil IRAP. Rapperswill

Špaček M (2015) Multilevel governance for balanced development between core and peripheral spaces. České vysoké učení technické, Praha, pp 7–15

The Guardian (2012) David Cameron to unveil year-long relaxation of planning laws. https://www.theguardian.com/politics/2012/sep/06/david-cameron-planning-laws-growth-plan. https://www.theguardian.com/politics/2012/sep/06/david-cameron-planning-laws-growth-plan. Accessed 19 Feb 2018

Zajko M, Brighton D (2015) Conceptual frameworks for higher education institution and industry collaboration. In: Brighton D, Zajko M, Pezoldt K (eds) Successful innovations? Efficient knowledge and technology transfer and international collaboration. Universitätsverlag Ilmenau, Ilmenau

Chapter 3
Spatial Planning Policies and the Integration Models as a Means for a Better Delivery of Services of General Interest

Špela Kolarič, Barbara Černič Mali, and Naja Marot

3.1 Introduction

Lately, several reports show that member states will struggle to implement the European Union's (EU) policy ambitions such as Territorial Agenda of the European Union 2020 and Commission White Paper on Services of General Interest (shorter SGI) to provide quality SGI[1] everywhere within the union (ESPON 2013a; Clifton et al. 2016; Gruber et al. 2017). According to the ESPON (2013a) report, the provision of SGI is mainly a domain of member states at national, regional, and local level, since aspects such as a minimum level of availability, accessibility, affordability, quality, and variety of a specific SGI are not determined by the EU legislation. Therefore, the impact of EU's policies on the implementation and delivery of SGI is fairly weak except for the regulations regarding the common market. The management and the administration of services is an issue to be dealt with foremost by providers of SGI, and new concepts like integration have been sought for. The integrated approach to the SGI provision is especially important for the remote areas due to their characteristics, such as dispersed settlements, dynamic and steep terrain, landscape change (Soliva 2007), demographic change—decreasing and ageing population (Gretter et al. 2017; Ferrario and Price 2014), and a challenging infrastructure construction (e.g. transport, telecommunication), which make the supply difficult and often

[1] The SGI definition is adopted by Gløersen et al. (2016), who divided SGI into services of general economic interest (SGEI), non-economic services (NSGI), and social services of general interest (SSGI), according to the nature of their provision. This is determined by who is the provider of the service (public/private/NGOs and social enterprises) and how is the service delivered to the users (are the prices market or state regulated, is the service free or subsidised, etc.).

Š. Kolarič · B. Č. Mali · N. Marot (✉)
Department of Landscape Architecture, Biotechnical Faculty, University of Ljubljana, Ljubljana, Slovenia
e-mail: naja.marot@bf.uni-lj.si

© Springer International Publishing AG, part of Springer Nature 2019

M. Finka et al. (eds.), *The Role of Public Sector in Local Economic and Territorial Development*, EAI/Springer Innovations in Communication and Computing, https://doi.org/10.1007/978-3-319-93575-1_3

economically unviable. Several studies (Pearce et al. 2005; Egger et al. 2011; Arcelus et al. 2015; Giuliani et al. 2017) have shown that remote areas are facing significant decrease of services, especially in regard to the basic goods provision (grocery stores, post and bank offices, pharmacies, doctor's offices). In addition to the closures, new forms of provision, predominantly smart e-services (Mayer et al. 2008; Dijkstra et al. 2014; Brozzi et al. 2015; Cledou et al. 2018; Khan et al. 2017) are replacing the former provision on location which might not be used to the full potential due to the predominantly elderly population and poor broadband coverage of the remote areas (Giuliani et al. 2017). In regard to these smart solutions, the tasks of the public sector have been adapted and today we talk more about the IC technology than place-based solutions. However, the later austerity situation in public services calls for further adaptation of the SGI provision, as in the UK example presented in Pinch and Patterson (2000)—for reorganisation and better co-operation among the sectors.

Integration, as a rather general "policy" term, has gained attention in the planning in the last 10 years (most references are from 2005 and later) and is in the multidisciplinary field best related to the environmental planning and policy analysis (Weingarten 2010; Fischer et al. 2013; Tajima and Fischer 2013; Eum et al. 2013). The understanding and conceptualisation of the integrative spatial planning can be best related to the communicative (Innes and Booher 1999, 2003, 2004), participative (Healey 1999, 2006), and co-evolutionary planning (Boelens and de Roo 2016; Mees et al. 2016), that all put emphasis on linking various stakeholders in planning with the purpose of establishing a dialogue for developing joint solutions in the communicative manner. On the basis of several authors and their view on the integration and its conceptualisation (Sykes et al. 2010; Stead and Cotella 2011; Lord 2013; Waterhout et al. 2012; Stanley 2014), we could derive the following three meanings of integrative spatial planning:

1. The integration of approaches such as strategic spatial planning, public participation processes, co-creation, research by design, impact assessment, and evaluation methods
2. The integration of themes, including resilience, climate change, (renewable) energy (production), and landscape
3. The integration of stakeholders in vertical (between administrative levels) and horizontal axes (between different sectors) in purpose-defined, trans-boundary soft spaces

With no general consensus on the definition of integration within the planning theory, we have based our understanding of the integration on the interpretations by Healey (2006), and Lloyd and Peel (2005). Healey (2006) explains it as a concept of four overlapping dimensions: (1) the (co)aligning of strategies and policy, (2) policy (re)framing, (3) connection between policy and action (policy and implementation), and (4) co-operation among actors. Lloyd and Peel (2005) provide a more detailed explanation and connect integration with different governance elements such as linking actors and sharing knowledge: "*Integration can imply co-ordinating strategy-making to avoid conflicting policies and to generate win-win situations. It might also imply broadening a policy frame to encompass a new issue. It can also suggest closing implementation deficits that can arise between policy and action. Finally, it implies linking actors together, sharing and developing knowledge for mutual benefit, often to overcome a fragmentation of institutional environments or a need to enter partnerships to achieve common goals. It is recognized that in pursuing integration, the ways in which it works in practice will be determined by local cultural practices and path-dependent factors* (Lloyd and Peel 2005)".

Based on these interpretations, we have formed a definition of the integrated territorial strategy, which is: "*a strategy for the SGI provision based on the four main principles: quality, availability, affordability, and accessibility, which takes into account peoples' actual and future needs in a given territory, territorial dimension, and the benefits of the synergies between the different SGI sectors* (Marot and Damjanovič 2016, p. 21)".

Similarly to the term's definition, the literature review revealed there is no commonly accepted definition of the integration models. Different studies, mostly performed in the frame of ESPON (2013a, b), have formulated individual definitions. Gløersen et al. (2016) describe the two single-element models (a) horizontal integration of actors during policy and programme preparation and (b) the integration of the financial instruments), and one combined model (c-integrating finance and policies) of integration, which lately occur in relation to the SGI provision. From this, a basic definition of the integration models was defined as the integration of either (1) actors, or any other identified elements: (2) policies, (3) administrative levels, (4) financial sources, and (5) other integration (e.g. of services, measures, sectors). Therefore, the purpose of this chapter is to reveal the level to which integration is already presented in the existing policy documents and legislation and what innovative solutions for the SGI delivery have been developed within the public sector policy framework. This is important in order to evaluate to what extent France, Italy (coverage of the Lombardy Region and South Tyrol), Austria (Tyrol, Carinthia), Switzerland, and Slovenia possess grounds for the introduction of the integration and thus innovative solutions for the delivery of SGI. Furthermore, this policy analysis served as validation of the integration and integration models' conceptualisation.

3.2 Approach to the Policy Analysis

The information included in this chapter results from the INTESI[2]EU project. The project with the title "Integrated Territorial Strategies for the Services of General Interest" has started in December 2015 in the frame of the Alpine Space programme, joining 10 partners from the state/cantonal/province level, research, associations, and services provision from France, Italy, Austria, Switzerland, and Slovenia. All partners contributed equally to the policy analysis, comparing five

[2] For more, please check: www.alpine-space.eu/projects/intesi/en/home *(quoted February 15th, 2018).*

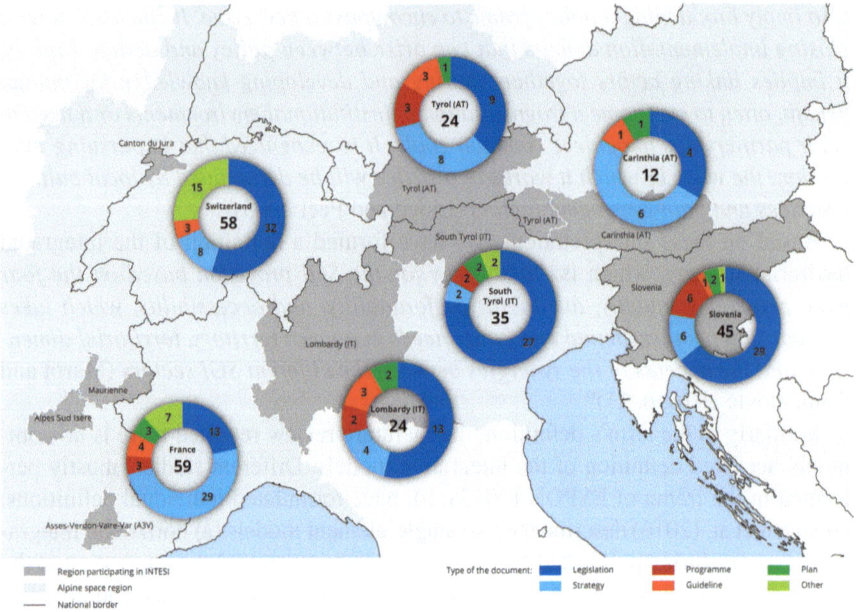

Fig. 3.1 Types of the documents submitted in the database in numbers by regions (Graphic by Nejc Florjanc and Barbara Kostanjšek)

participating countries and their regions (Austrian Tyrol and Carinthia, Italian Lombardy and South Tyrol, French Auvergne Rhône-Alpes, Slovenia, and Swiss Canton du Jura), and the eight examined sectors (general, regional development, transport, telecommunication, basic goods, health, social care, and education). The data for the comparison analysis (Kolarič et al. 2017) was retrieved from the following project activities:

(a) the Database of the Existing Strategies
(b) Seven Regional Reports on the policy documents and legislation governing the selected eight SGI sectors: Tyrol (AT), Carinthia (AT), Lombardy (IT), South Tyrol (IT), Auvergne Rhône-Alpes (FRA), Slovenia (SI), and Canton du Jura (CH)
(c) the Transnational Workshop (June 2016)

The database, which is publicly available on the project website, includes 257 documents altogether, the geographical coverage and types of the documents are visible from Fig. 3.1. The information it contains on the reviewed documents was mainly used to conduct the numerical analysis in Excel. The documents listed are described with 18 basic categories (e.g. name of the country, sector, original and English name of the document, administrative level, type, year of adoption, duration, and major objectives) that were used to conduct the numerical and qualitative analysis.

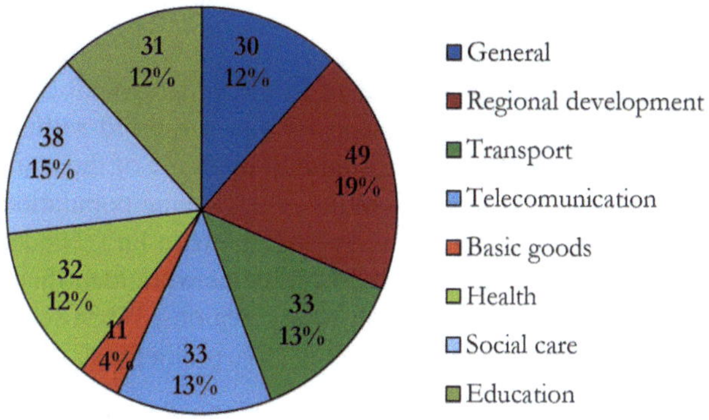

Fig. 3.2 Overall number and share of the documents by sectors

When making the selection, the countries aimed for a balanced representation of the documents by administrative levels (national, regional, local), and inspected sectors. The overview of the submitted documents (Fig. 3.2) shows a rather equal division of around 30 entries per sector (12–15%) with the regional development sector as the most represented one (Kolarič et al. 2016). The majority of the documents by type are legislations (49%), followed by strategies (25%), guidelines (6%), programmes (5%), and plans (5%).

The comparison analysis is based on seven regional reports: Tyrol (AT), Carinthia (AT), Lombardy (IT), South Tyrol (IT), Auvergne Rhône-Alpes (FRA), Slovenia (SI), and Canton du Jura (CH). The regional reports among others summarise the information gathered by altogether 70 interviews, from 8 to 13 in each observed region. The interviews were conducted by the PPs in their national languages to support the review of the selected strategies for the SGI provision and collect the information on responsible stakeholders, governance models, implementation process, measures, efficiency, and ways of financing. The PPs had the liberty to modify (add/remove) some of the 12 questions to a certain extend as needed. The interviewees were the stakeholders (e.g. mayors, public officials, services providers) from various governance levels (state, regional, provincial, local), identified by the PPs as important for the provision and governance of SGI in their TAs or broadly.

The information gathered at the Transnational Workshop conducted on 29th June 2016 also served as an input to the analysis. It involved 19 attendees who were divided by countries into five groups (Austria, France, Italy, Slovenia, and Switzerland). The aim of the workshop was to gather and compare the information on: (1) the understanding of the SGI definition and the concept of integration, and (2) the presence or absence of integrated SGI strategies in the participating countries. Additionally, the existing governance models that support the integrated SGI delivery were identified for all five countries.

3.3 Problems of the SGI Delivery

The study shows general problems related to the geographical characteristics of the regions such as a dynamic and steep terrain, dispersed settlements, scares, decreasing and ageing population, are common to all (7) of the analysed regions. The only region not tackling the difficulties of the ageing population and furthermore, reporting a population increase is Swiss Canton du Jura (Niederer and Egger 2016). Reviewing the regional reports, 14 problems were identified altogether in relation to the SGI delivery in the remote alpine regions. When selecting the ones present in at least three of the analysed regions, five problems stand out (Table 3.1):

1. Accessibility of SGIs in mountainous regions highly depends on the spatial location and the quality of public transport.
2. Underdevelopment of the infrastructure and services in terms of poor quality and/or supply.
3. Costs (expensive delivery, austerity, no allocated funds, no investments, etc.) of the services in Alpine areas.
4. Unresponsiveness of the governance system to the actual (changing) needs of the local communities.
5. The strategies are too broad and often lack the specification of measures (how to do something), responsibilities (who should do it), and monitoring mechanisms.

The challenges related to accessibility, clearly demand the integration of individual services with the transportation. In relation to the high cost (problem 3), the integration of services would enable a more efficient use of financial sources.

Table 3.1 Problems of the SGI delivery in the Alpine regions

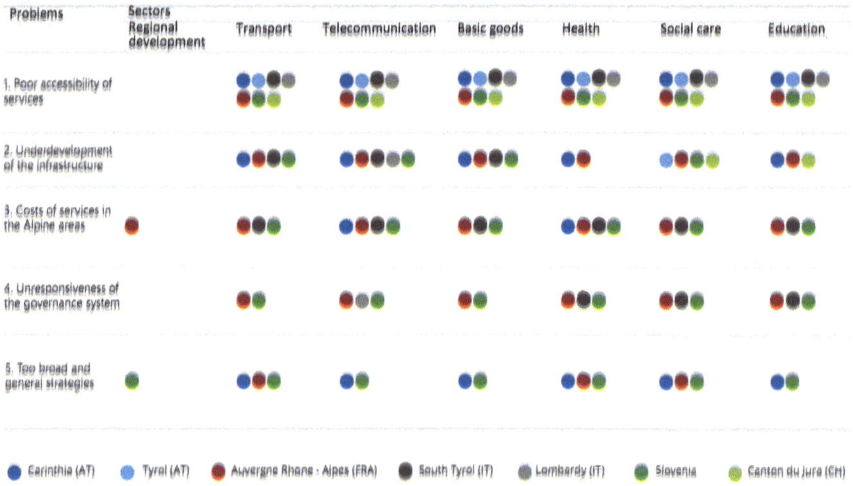

If the PPs did not specify in their reports to which specific sector the problem that they have identified and listed refers to, it was considered that the issue is present in all the inspected sectors

Unresponsiveness of the governance system was reported for the French, Slovenian, and Italian regions, while inadequate strategic policies were also identified in Austria. The lack of specification of measures, responsibilities, and monitoring mechanisms could be addressed by a better integration of actors, administrative levels and policies which altogether indicates the need for the transformation of the public sector's role.

3.4 Integration

The Presence of Integration

Using the information in the database, the analysis of all the 257 documents shows, more than half (64%) include the integration as a concept. However, looking at the number and percentage of the analysed documents, the best coverage was evaluated by Italy and France (90%) and the worst by Switzerland. The Austrian documents comprise the integration in 78%, and the Slovenian in 67%.

As shown in Fig. 3.3, the highest percentage of the documents that comprise the integration belongs to the regional development sector (76%), closely followed by the documents referred as "general". Most likely this is due to the fact that regional and general policies serve as umbrella policies and thus should be integrative per se. The basic goods, transport, telecommunication, and education sectors all include

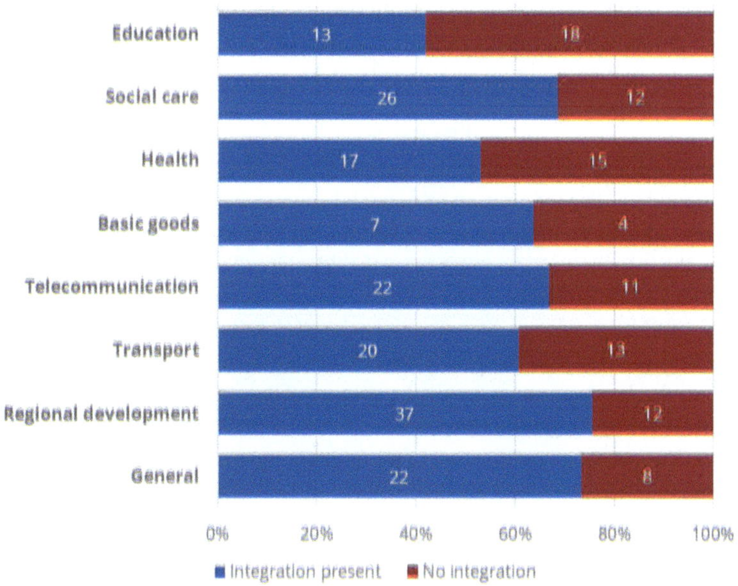

Fig. 3.3 The relativised number of the analysed documents including the integration by sectors

the integration in roughly 60%, whereas the health care documents include it in only 47%. The integration as a concept is at least present in the social care sector (32%) though this is one of the sectors that together with the health requires significant changes in the SGI provision models due to the ongoing ageing of the population. Also, in this sector public providers are no longer able to assure sufficient services and increasingly rely on supply from the private sector.

Concerning the distribution of the integration in the sectoral documents, the results of the transnational workshop (Marot and Damjanovič 2016) show a very similar picture. According to the participants, integration is mostly present in the strategic regional development documents, followed by the telecommunication, transport, general sector, and health sector. Again, a lack of integration in the social care sector was confirmed. The workshop's participants also stated the integration is missing in the sector of basic goods, whereas the database analysis shows the concept is present in 64% (7 out of 11 documents). However, in all the countries together, only 11 documents have been selected for this sector. This implies that the provision of basic goods is not managed by the strategic SGI policies and is not primarily in domain of public sector, but is mainly a subject to the market conditions.

Existing Integration Models

The integration models were according to the planning system element they integrate defined into five categories: actors, policies, administrative levels, financial sources, and others. According to a more detailed analysis, actor-based integration models (e.g. national authorities, regions, municipalities, interested public, service providers) are prevailing in all sectors (Fig. 3.4).

The integration among the administrative levels (e.g. national, regional, municipal, local) in which actors on different levels communicate and co-operate is best represented in the general, transport, and health sector, is also one of the more often occurring models. The integration of financial sources is more common in the health, social care, education, and regional development sectors. However, as further confirmed by the interviewees, in most cases the "declared" integration of actors, administrative levels, or finances, presented in the documents, does not indicate the integrative and "intersectoral" approach to the SGI delivery in practice. That is also supported by the fact that the policy and other (e.g. services or sectors) integration models are the least common in all the sectors. Therefore, one can question if the "declared" level of integration corresponds to the actual "implemented" integration.

As one document can predict two or more integration models (e.g. actors, finances, and policies), a more detailed analysis looked at the combinations of the recognised integration models in the analysed documents. Altogether, 162 combinations of the 5 integration models or elements (actors, policies, administrative levels, financial sources, and other) were identified which reveals a complex nature of the integrative approach. The documents comprise 117 different "combined" (when at least two

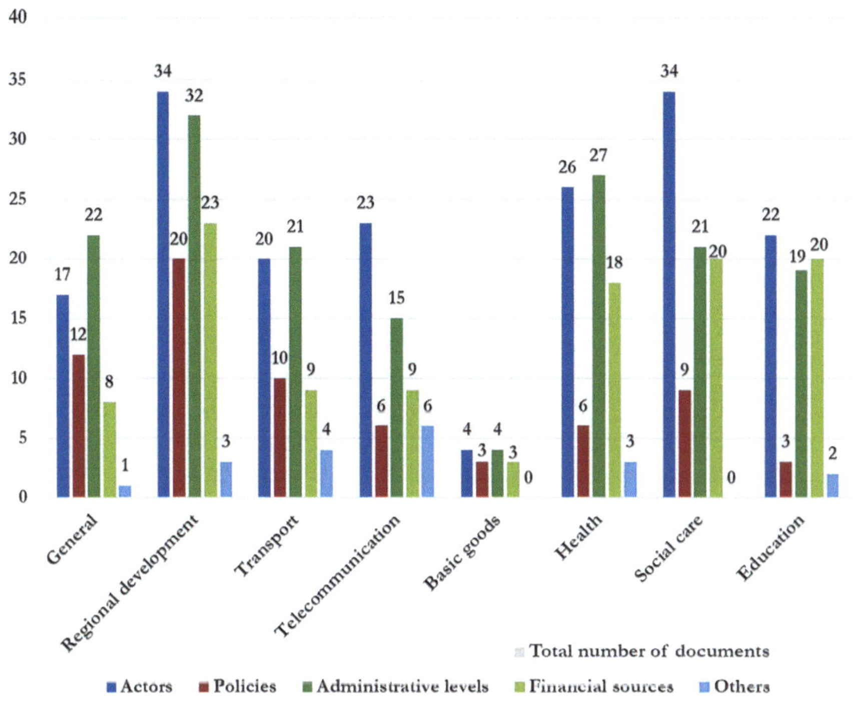

Fig. 3.4 The number of the occurrences of the integration models in the documents by sectors

integration models are predicted—e.g. actors and finances) models, which largely prevail over the 45 "singular" (when only one integration model is predicted—e.g. integration of actors) models of integration. However, looking at the various combinations, the singular models (comprising 1 element) are next to the models combining two elements (45), the most common, followed by the once combining three (41).

The combination of all the five models (5E models) was noted twice, once in the transportation and once in education sector. In the models with four elements (4E models), the combination of actors, policies, administrative levels, and finances is the most common (altogether 24 examples present in all the sectors except of general and basic goods). In the models comprising three elements (3E models), two combinations stand out. One combines actors, administrative levels, and finance (17 cases in all sectors, apart from the general and basic goods), and the other actors, policies, and administrative levels (16 examples, all sectors, apart from education). These two models represent the vertical integration which optimises the delivery of SGI along the vertical administrative axes and also corresponds to the subsidiary principle. In the two-element (2E) models, the two obvious combinations are of the actors and administrative levels, and the administrative levels and finances. The most common among the singular models is the one integrating actors (20 examples, most in the social care sector), followed by the integration of the administrative levels (13 examples, mostly in the general and regional development sector).

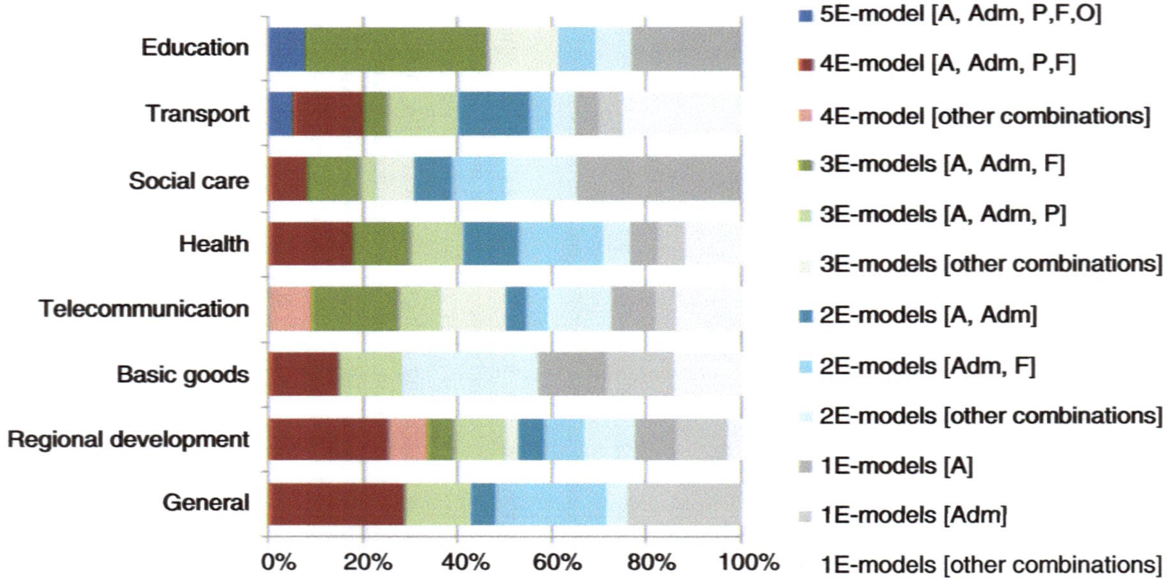

Elements key: A - actors, Adm - administratiom level, P - policics, F - financial sources, O - others
Models key: e.g. 5E-model - 5 (number of elements) E (abbreviation for element)

Fig. 3.5 Division of the combined and singular models of integration by sectors

Figure 3.5 illustrates that the most complex models of integration are present in the general and regional development sectors. The models reflect the complexity of the sectors in terms of their overarching, guiding, and connecting (actors, policies, sectors) features. More simple integration solutions are present in the sectors of social care, health, basic goods, and telecommunication, in which actors and administrative levels, or administrative levels and finances are joined. As solutions for these sectors, mostly practically implemented integration models are required, as oppose to the policy-oriented models in general and regional development. In the education sector, the 3E-model connecting actors, administrative levels, and finance stands out. In the social care sector, the integration of actors is predominant, connecting the social care services, home care assistance, municipalities, homes for the elderly, etc.

Examples of the Integration Models in Practice

Integration models in practice can be considered as innovative solutions for the regions, and are in most cases also targeting the services provided by the public sector. Regions report these examples as either "integration models", "cases", or "examples of integration". As shown in Table 3.2, most of the existing integration models linking the governance elements (actors and administrative levels) do not mention the sector integration. An exception to that is the Tyrolian exploitation of synergies among the ongoing civil engineering projects when expanding broadband.

In addition to the policy-related models, significantly more cases were catalogued as models in practice. As seen from Table 3.3, the services integrate two or more sectors (e.g. provision of meals for school children and elderly in co-operation with local producers). The most often combined are the transportation sector with either education, health, or social care, and the telecommunication sector with health and social care which indicates the introduction of the ICT technological solutions to the SGI provision. At the most, four different sectors are integrated in one service. Examples of such services are: the Regional Card for Services (RCCS), introduced by the Region of Lombardy (integrating administrative, telecommunication, health and social care sectors and services; Pellegrino et al. 2016), Roaming Services which are in use in France (including transport, education, social care, and basic goods; Devos et al. 2016), and the Provision of meals for school children and elderly in co-operation with local producers—a model implemented in the Italian South Tyrol (Giuliani and Hoffmann 2016).

The services integrating three sectors are: the Transport services for the out of school activities by students and youth (South Tyrol-IT), the Mobile health and social services for elderly (both present in South Tyrol-IT; Giuliani and Hoffmann 2016), and a Postbus (it. Auto-postale), integrating transport of people, medicines, and goods, which is in place in Lombardy. A Digital ecosystem E015 is a service also introduced by Lombardy, which allows for the communication among the IT services of public and private actors operating in multiple sectors: transport, hospitality, tourism, culture, etc. (Pellegrino et al. 2016). There are other integration mod-

Table 3.2 Examples of the existing governance integration models in the participating regions

Examples	Type of the integration model		
	Actors	Adm. levels	Policies
STRALE!K (Carinthia, AT): is a strategic rural development programme. It addresses all topics of the SGI and defines individual solution approaches for their maintenance. It provides of guidelines and objectives of the SGI delivery. It is a base for various department strategies. There is a special governmental office dedicated to the implementation of this programme which implies a new organisation unit of public sector			X
Establishment of regional management (Tyrol, AT): Established for the implementation of regional strategies/concept/programmes. Members are the municipalities, tourism associations, district chambers, and interested companies from the region. The aim is to pursue a target-oriented co-operation of the regions with the provincial government, federal government, and the EU, and promote the strategic regional development	X	X	
Synergies among civil engineering projects (Tyrol, AT): Broadband infrastructure expansion is conducted at the local level by exploiting synergies in the course of civil engineering projects conducted by public or public-owned carriers (gas network, district heating network, sanitary engineering, road construction, etc.)	X		
Social entrepreneurship (Slovenia): A promising future model for integration and provision of mostly social services (child care, elderly acre, social inclusion, etc.), funds, and stakeholders	X		X
Service plans (Auvergne Rhône-Alpes, FRA): An integrated territorial approach model to SGI delivery. By law French departments (administrative level similar to municipalities) must have their Departmental Plan of Improved Accessibility to Services of General Interest (SDAASP), which calls for shared services governance, and coordination and innovation in the provision of services, taking into account the specificities of each territory	X		X

els such as a Mountain Virtual Hospital (MVH), an experimental model of mountain hospital (introduced by the region of Lombardy within Valchiavenna Area Strategy), which uses innovative services and new technologies, such as tele-medicine, home-based tele-radiology, points of care, and de-localised diagnostic treatment-rooms (Pellegrino et al. 2016).

Innovative aspects of these models are shared services of governance, new organisational units, restructuring of the service provision and change in the responsibilities of public sector, introduction of IC technologies for the provision of SGI, and newly established co-operation models between sectors, e.g. telecommunication and health, or transport, education, and health.

Table 3.3 Existing examples of services integration in the participating regions

Integration models examples	Region	General	Transport	Telecomm.	Education	Health	Social care	Basic goods
Regional card for services (RCS) – multi function smart card (access to public administration, health, social services)	Lombardy (IT)	X		X		X	X	
Integrated ticket for different modes of transport for users and pupils	Slovenia		X		X			
TravelPlanner webportal aiming at gathering train, bus, metro timetables and routes	Lombardy (IT)		X	X				
Roaming services: libraries, cinema, basic goods, pedicure…	Auvergne Rhone - Alpes (FRA)		X		X		X	X
Mobility card: young & museums	South Tyrol (IT)		X		X			
Transport services for extra activities by students and youth	South Tyrol (IT)		X		X		X	
A postbus (Auto-postale) integrating transport of people, medicines, goods	Lombardy (IT)		X			X		X
Healthcare card enabling management of medical services	South Tyrol (IT)			X		X		
Provision of meals for school children and the elderly in cooperation with local producers	South Tyrol (IT)				X	X	X	X
Multidisciplinary health houses/ territorial hospital groups	Auvergne Rhone - Alpes (FRA)					X		
Mobile health & social services for elderly people	South Tyrol (IT)		X			X	X	
Social concept of Ausserfern – mobile care givers	Tyrol (AT)		X				X	
Regional Health Care Structure Plan - Outpatient Module – transition care after the acute care at the hospital	Tyrol (AT)					X	X	
Info points for tourists at small retailers	South Tyrol (IT)	X						X
Consumer cooperative KonsuMoos for basic goods (municipality/inhabitants)	South Tyrol (IT)	X						X
Wanderhandel - mobile bakery, dairy products	South Tyrol (IT)		X					X
Tourist or guest card - card allowing access to a variety of services		X		X			X	
Mountain Virtual Hospital (MVH), experimental model of Mountain Hospital using innovative services and technologies				X		X		
Digital ecosystem E015 enabling IT communication among multiple public and private actors and sectors: transport, hospitality, tourism, culture, etc.			X	X			X	

Carinthia (AT) · Tyrol (AT) · Auvergne Rhone - Alpes (FRA)
South Tyrol (IT) · Lombardy (IT) · Slovenia

3.5 Discussion with Conclusion

The performed analysis can lead to several SGI delivery, planning theory, and policy-making conclusions. Firstly, the identified gaps in SGI delivery across the Alps showed that while some barriers for the SGI delivery can be predicted due to the physical and socio-geographical characteristics of the areas, there are other governance-related issues that need to be addressed. Additionally, economic crisis has in addition to the financial burdens also weakened the public sector concerning human capacities. The latest administrative changes in France and Austria demanded reorganisation as well in service delivery or some of the services have been due to inefficiency simply shut down. In these conditions, the public sector is looking for new ways to deliver and fulfil the basic needs of the population through several innovative approaches. This chapter focused on integration as planning approach, and from the results the following limitations and strengths of the approach can be depicted.

The tested definition of the integration deemed applicable and could be further disseminated though from the presented cases integration starts only if at least two or three governance elements are included. From the sectoral view we can also talk about the integration within the sector, most common among transportation sectors. This comes from a discussion on how to diversify between the sectoral and service integration, which concluded that each sector provides multiple services, so the integration can also happen within the sector between two services. The most complex models of integration are present in the general and regional development documents, as these sectors are overarching, guiding, and connecting different policies, actors, sectors, etc. More simple, practice-oriented integration solutions are present in the sectors of social care, health, basic goods, and telecommunication. These solutions usually demand just simple accommodation of the current practises of public sector and mostly demand increased engagement and co-operation of actors without larger financial investments.

In terms of sectoral integration, the transport sector should be better linked with telecommunication, education, health, social care, and basic goods. Furthermore, the services within the sector itself need to be connected which is currently addressed with several mobility initiatives across Europe. In the times of the "digital society", the integration of the telecommunication services with other sectors (health, social care, education, transport, and basic goods) is also recognised as important and already seems to override the importance of place-based solutions in regard to the SGI delivery. The information communication technology (ICT) represents an opportunity for the (economic, social, etc.) development of Alpine areas, as it could bridge the challenges related to "physical" remoteness and the hindered accessibility, by enabling the SGI services online.

Regarding the policy-making, it is necessary to emphasise the difference between the declared and implemented integration. The interviewees explain the "declared" integration recognised in the documents, is not necessarily transferred into an integrative approach to the SGI delivery in practice. This is further confirmed by the analysis showing the integration of policies and other elements (e.g. sectors, services, measures) in practice is seldom. To improve the accessibility and supply of SGIs in the Alpine regions, public sector should better invest resources into better linking primarily transportation and telecommunication services need to be linked with other sectors. Additionally, identified practised models of SGI integration should be evaluated in a transnational framework for their utilisation potential across Europe. Therefore, the policy-makers should ensure that the new spatial planning policies offer comprehensive solutions and reflect these needs by encouraging and enabling the integration models. Some pre-requisitions for the integration are already defined in the legislation, such as subsidiary principle in the planning legislation which demands from the lower administration levels to accommodate to the provision on the higher administrative levels. This could also be achieved by flowing the principles of communicative planning and with the development and application of the participative planning tools (e.g. round tables, workshops, focus groups). However, the success of the integrated SGI models largely depends on the far-reaching power of the planning policies in the inspected regions. As shown, the general and regional policies already present a good base for the SGI integration, which they at the moment fail to deliver. An interesting observation of the analysis is also that the public sector has reservations about using the new ways and models of the SGI delivery (e.g. the available digital tools), which might also explain the difference between the declared and actual integration to some extent. Better dissemination and training will be required in the future in order to adopt the SGI delivery as supported by the public sector to the current needs and economic conditions.

References

Arcelus FJ, Arocena P, Cabasés F, Pascual P (2015) On the cost-efficiency of service delivery in small municipalities. Reg Stud 49(9):1469–1480

Boelens L, de Roo G (2016) Planning of undefined becoming: first encounters of planners beyond the plan. Plann Theory 15(1):42–67

Brozzi R, Lapuh L, Nared J, Streifeneder T (2015) Towards more resilient economies in Alpine regions. Acta Geogr Slov 55(2):339–350

Cledou G, Estevez E, Barbosa LS (2018) A taxonomy for planning and designing smart mobility services. Gov Inf Q 35(1):61–76

Clifton J, Díaz-Fuentes D, Fernández-Gutiérrez M (2016) Public infrastructure services in the European Union: challenges for territorial cohesion. Reg Stud 50(2):358–373

Devos A, Horgues-Deba J, Daudé R, Doukhan G (2016) Regional Report Auvergne Rhône-Alpes France. INTESI—Integrated Territorial Strategies for Services of General Interest/2015–2018. Association pour le Développementen Réseau des Territoireset des Services, Gap/Chambéry

Dijkstra L, Garcilazo E, McCann P (2014) The effects of the global financial crisis on European regions and cities. J Econ Geogr 15(5):935–949

Egger T, Niederer P, Falempin L, Becker C, Stephan C (2011) Strategies to improve accessibility to SGI in rural mountain areas. Final synthesis of the Interreg IVB ACCESS project, SAB, Berne

ESPON (2013a) SeGI Indicators and perspectives for services of general interest in territorial cohesion and development. Applied Research 2013/1/16.Final Report/Version 25/05/2013 Executive Summary (Rep.). ESPON & Royal Institute of Technology

ESPON (2013b) TANGO-Territorial Approaches for New Governance, Executive Summary. Applied Research 2013/1/21.Version 20/12/2013 Executive Summary (Rep.). ESPON & Royal Institute of Technology

Eum JH, Scherer D, Fehrenbach U, Köppel J, Woo JH (2013) Integrating urban climate into urban master plans using spatially distributed information. Land Use Policy 34:223–232

Ferrario E, Price M (2014) Should I stay or should I go. Alpine brain drain and brain gain: the reasons behind the choices of young mountain people. J Alp Res. https://doi.org/10.4000/rga.2381

Fischer TB, Smith M, Sykes O (2013) Can less sometimes be more?—Integrating Land Use and Transport Planning on Merseyside (1965–2008). Urban Plann Transp Res 1(1):1–27

Giuliani C, Hoffmann C (2016). Regional Report South Tyrol—Italy. INTESI—Integrated Territorial Strategies for Services of General Interest /2015–2018. European Academy of Bolzano Institute for Regional Development and Location Management, Bolzano

Giuliani C, Hoffman C, Laner P (2017) WPT2 Assessment Regional Report. Deliverable 3.7.2 EURAC Research. Synthesis Report. EURAC Research, Bolzano, 69p

Gløersen E, Drăgulin M, Haarich S, Zillmer S, Holstein F, Lüer C, Hans S (2016) Research for REGI Committee—Services of General Interest in the Funding Period 2014–2020 (Study). European Parliament, Directorate-General for Internal Policies

Gretter A, MacHold I, Membretti A, Dax T (2017) Pathways of immigration in the Alps and Carpathians: social innovation and the creation of a welcoming culture. Mt Res Dev 37(4):396–405

Gruber E, Rauhut D, Humer A (2017) Territorial cohesion under pressure? Welfare policy and planning responses in Austrian and Swedish Peripheries. Pap Reg Sci. https://doi.org/10.1111/pirs.12344

Healey P (1999) Institutionalist analysis, communicative planning, and shaping places. J Plan Educ Res 19(2):111–121

Healey P (2006) Territory, integration and spatial planning. In: Tewdwr-Jones M, Allmendinger P (eds) Territory, identity and spatial planning: spatial governance in a fragmented nation. Routledge, London, pp 64–79

Innes JE, Booher DE (1999) Consensus building and complex adaptive systems: a framework for evaluating collaborative planning. J Am Plan Assoc 65(4):412–423

Innes JE, Booher DE (2003) Collaborative policymaking: governance through dialogue. In: Hajer MA, Wagenaar H (eds) Deliberative policy analysis: understanding governance in the network society. Cambridge University Press, Cambridge, pp 33–59

Innes JE, Booher DE (2004) Reframing public participation: strategies for the 21st century. Plann Theory Pract 5(4):419–436

Khan Z, Pervez Z, Abbasi AG (2017) Towards a secure service provisioning framework in a Smart city environment. Futur Gener Comput Syst 77:112–135

Kolarič Š, Marot N, Černič Mali B (2016) Regional Collection of Strategies. INTESI—Integrated Territorial Strategies for Services of General Interest/2015–2018. University of Ljubljana, Ljubljana

Kolarič Š, Marot N, Černič Mali B, Kostanjšek B (2017) WPT1 Deliverable 1.2.3 Report on Comparison Analysis. EURAC Research, Bolzano, 90p

Lloyd G, Peel D (2005) Tracing a spatial turn in planning practice in Scotland. Plan Pract Res 20(3):313–325

Lord AD (2013) Towards a non-theoretical understanding of planning. Plann Theory 13(1):26–43

Marot N, Damjanovič V (2016) Report on Transnational Workshop. INTESI—Integrated Territorial Strategies for Services of General Interest/2015–2018. University of Ljubljana, Ljubljana

Mayer M, Woltering M, Hubert J (2008) Tourism and regional development in the Bavarian Alps. Geogr Rundsch 60(10):40–46

Mees H, Tempels B, Crabbé A, Boelens L (2016) Shifting public-private responsibilities in Flemish flood risk management. Towards a co-evolutionary approach. Land Use Policy 57:23–33

Niederer P, Egger T (2016) Regional Report Canton du Jura—Switzerland. INTESI—Integrated Territorial Strategies for Services of General Interest/2015–2018. Schweizerische Arbeitsgemeinschaft für die Berggebiete, Bern
Pearce G, Ayres S, Tricker M (2005) Decentralisation and devolution to the English regions: assessing the implications for rural policy and delivery. J Rural Stud 21(2):197–212
Pellegrino C, Cappiello A, Lentini G (2016) Regional Report Lombardy—Italy. INTESI—Integrated Territorial Strategies for Services of General Interest/2015–2018. RegioneLombardia, General Directorate University, Research and Open Innovation, Milano
Pinch PL, Patterson A (2000) Public sector restructuring and regional development: the impact of compulsory competitive tendering in the UK. Reg Stud 34(3):265–275
Soliva R (2007) Agricultural decline, landscape change, and outmigration: debating the sustainability of three scenarios for a Swiss mountain region. Mt Res Dev 27(2):124–129
Stanley JK (2014) Land use/transport integration: starting at the right place. Res Transp Econ 48:381–388
Stead D, Cotella G (2011) Differential Europe: domestic actors and their role in shaping spatial planning systems. disP 186(3):13–20
Sykes O, Lord A, Thakur J (2010) Planning in a World-Container. Town and Country Planning, January 2010
Tajima R, Fischer TB (2013) Should different impact assessment instruments be integrated? Evidence from English spatial planning. Environ Impact Assess Rev 41:29–37
Waterhout B, Othengrafen F, Sykes O (2012) Neo-liberalization processes and spatial planning in Northwest Europe: an exploration. Plan Pract Res 28(1):141–159
Weingarten E (2010) Merits of a more integrated approach to environmental assessments. Environ Policy Gov 20(1):12–29

Chapter 4
Utilising Endogenous Potentials via Regional Policy-Led Development Initiatives in (Post-) Industrial Regions of Central Europe

Jörn Harfst, Peter Wirth, and Danko Simić

4.1 Introduction

Speaking about the role of the public sector in local economic and territorial development, one has to consider the opportunities as well as the limits of policymaking at the lower levels of multi-level governance arrangements. On the one hand, it is often stressed that local development can only be 'done' by local people (e.g. Gruenewald and Smith 2014). On the other hand, it is obvious that local and regional development highly depends on decisions on the European and national levels (e.g. Jeffery 2015; Pinder 2017). In this sense, this chapter intends to shed a deeper light onto the connection and role of different governance levels in the field of regional development by discussing the utilisation of space-based potentials in regional development.

Since the Barca Report to the European Union (Barca 2009) approaches to utilise endogenous potentials have been firmly on the European Union agenda, streamlined by various policies related to territorial development (e.g. European Union's Territorial Agenda). Endogenous territorial potentials are defined as being 'divided into tangible assets, such as natural and human resources, and intangible assets, like organisation, culture, social issues and governance' (Damsgaard et al. 2009: 9). At the end of the current EU programming period (2012–2020), it has to be discussed how such potentials are shaped and used, particularly in place-based development approaches at the local and regional level.

J. Harfst (✉) · D. Simić
Department of Geography and Regional Science, University of Graz, Graz, Austria
e-mail: joern.harfst@uni-graz.at; danko.simic@uni-graz.at

P. Wirth
Leibniz Institute of Ecological Urban and Regional Development (IOER), Dresden, Germany
e-mail: p.wirth@ioer.de

© Springer International Publishing AG, part of Springer Nature 2019
M. Finka et al. (eds.), *The Role of Public Sector in Local Economic and Territorial Development*, EAI/Springer Innovations in Communication and Computing, https://doi.org/10.1007/978-3-319-93575-1_4

In doing so, the chapter highlights the situation of areas characterised as non-agglomeration, (post-) industrial regions in Central Europe, often situated in the peripheries and outside the main academic focus. Cities and regions of this type often lag behind regions that have managed to transform towards spaces of knowledge-based services and updated industries. Many of these regions are undergoing a process of decline and shrinkage. Due to outmigration and ageing, they suffer from social erosion and a lack of human capacities. In consequence, many of these places have become 'chronic patients of regional policy, constantly in need of care but not getting well' (Wirth et al. 2016: 63).

The analysis will deal with the governance issues and the possible valorisation of endogenous development potentials by different policy levels. Whereas earlier work of the authors reflected on the inventory of endogenous potentials and different governance constellations in the affected regions (Harfst and Wirth 2011, 2014), this text discusses issues of agenda setting and implementation.

The cases presented in this chapter highlight the challenges in the current development of such regions, give examples for the identification and utilisation of endogenous potentials, and discuss some lessons learned from this locally led, place-based development approach. Additionally, different governance modes are discussed and used to illustrate different agenda setting in regard to endogenous potentials. As a result, the authors conclude that the utilisation of potentials is streamlined top-down via various policy frameworks, but additionally is also strongly shaped by local and regional governance settings, such as actors, networks and capacities.

4.2 Endogenous Potentials in Regional Development

In the academic discourse on regional development, endogenous potentials are usually considered from two different perspectives (Martin and Sunley 1998; Stough et al. 2011). Firstly, they are discussed in Economic Geography when analysing unbalanced growth and development patterns on different spatial levels (Dunford and Smith 2000; Petrakos et al. 2005). In this context endogenous potentials are seen as drivers of change (technology, research, human resources, etc.) displayed as an explanation of growth patterns—contrary to a neoclassical economic approach (Maier et al. 2006; Myrdal 1957; Romer 1994). Secondly, from the view point of applied regional sciences and regional development, endogenous potentials are used to explain the specifics of regional and local development patterns and approaches. There is a wide range of academic literature on 'industrial districts', 'clusters', 'institutional thickness' and 'learning regions' as proof of these discussions (among others, see Amin and Thrift 1995; Lazzeroni et al. 2013; Moulaert and Sekia 2003; Perlik 2001). These approaches all highlight regional factors for growth, such as institutions, capacities of actors, cultures and economic linkages within the regional or local context.

Especially against the backdrop of the financial and economic crisis after 2008, a strong dependency on global markets with a focus on foreign direct investments has proven a risky development strategy for many regions. Therefore it is not surprising that territorial and endogenous potentials are receiving renewed attention from many actors involved in regional development initiatives (Damsgaard et al. 2009).

Endogenous potentials also play a role in EU strategy papers, in particular in the Territorial Agenda 2020 (EU 2011). This builds on the Europe 2020 strategy's objectives (European Commission 2010), which set out concrete development goals for the European Union and its member states in terms of employment, research, climate change, education and poverty reduction. The Territorial Agenda 2020 assumes that these objectives can be achieved if their 'territorial dimension' is taken into account (EU 2011: 2).

The objectives of the EU—territorial and social cohesion and competitiveness—can only be reached if states, regions and cities (including small- and medium-sized cities) contribute to the European objectives (EU 2011: 3, paragraph 8). This requires local and regional development strategies based on 'territorial potential' (EU 2011: 3, paragraphs 11 and 28). Challenges include economic stability against external shocks, the effects of demographic (shrinkage, growth, ageing) and climate change, the use of renewable energies and the preservation of natural landscapes and cultural heritage. In order to solve these issues 'the potential of the respective territory should be raised' (EU 2011: 7). For the cities and regions, it is therefore important to break down the general requirements and to determine their potentials more precisely.

As a methodological base Damsgaard et al. (2009) use an approach that distinguishes between 'tangible' and 'intangible' potentials. The former is about concrete, material legacies of past human activity in settlements and landscapes. Regarding the (post-) industrial regions addressed here, one could think of mining dumps, old factory buildings, unused transport infrastructure and other features. On the other hand, it is also about buildings, facilities and infrastructures that date back to the industrial era. The second, immaterial type of potential relates to experiences, skills, knowledge and other competences, as well as cultural and social aspects of human existence. These non-material potentials are moreover an inner linkage between the people and the industrial past (identity) and cultural values, as well as traditions that shape the lives of the region's inhabitants (Fig. 4.1). Only, the utilisation of immaterial aspects is generally harder to operationalise, with only some projects having dealt with the utilisation of such potentials outside the tourism domain (Harfst and Simić 2017), while for example the conversion of old industrial sites is academically well documented and understood (e.g. Oevermann and Mieg 2014).

But it is also possible to search for completely different starting points that are not linked to the industrial past. Whether and how these potentials are valued depends largely on the preferences, perceptions and capacities of local and regional actors, but also on frameworks of European and national policies that can contribute to the construction of spatial constellations and challenges (Danielzyk 2012). The

Fig. 4.1 The interplay of agenda setting, endogenous potentials and implementation in regional development (source: authors' graphic)

challenge in utilising endogenous potentials in general is to discover or enhance the special, the specific, often unique, and the meaningful potentials in cities and regions and place them in the context of local or regional development.

In the following, the question how utilisation is shaped policy-wise on different levels is addressed, using two case studies of the exploitation of the specific endogenous potential of (post-) industrialised regions to generate positive momentum for development after a long process of structural change. Hereby the focus is more on showing the opportunities and limits of utilising endogenous potentials with particular attention paid to their policy setting, and less on characterising the 'problem type' of (post-) industrial regions as such.

4.3 Agenda Setting via Endogenous Potentials

As Princen (2007) points out, agenda-setting processes are crucial in urban development, since they determine which issues are taken up for decision-making and which are largely neglected. Therefore, such processes determine the range of legitimate concerns and alternatives in a political system. Formal agenda setting, as a part of the policy cycle within the EU, is firmly set with the European Commission and the European Parliament (Pollack 1997).

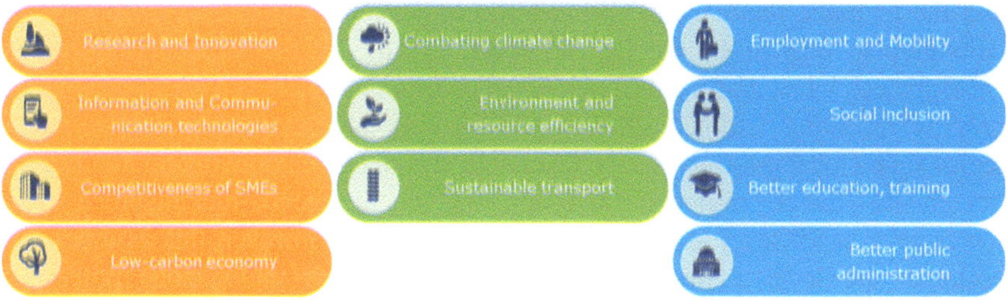

Fig. 4.2 Thematic fields of action according to the EU2020 goals (source: www.ec.europa.eu/regional_policy/en/policy/cooperation/european-territorial/)

When looking at the operationalisation of endogenous potentials on the political level, the role of the 'Territorial Agenda 2020' has been already highlighted in the previous section. Stemming from the recommendations of the Barca report, the approach towards the utilisation of endogenous growth potentials can be found in many sub-policies of the European Union's cohesion policy, e.g. several INTERREG programmes or the EU Commission's Smart Specialisation Initiatives, such as S3. Therefore, discussing policy arrangements and agenda setting around endogenous potentials requires a deeper look into multi-level governance setting between European, national and regional/local levels in order to comprehend how policy agendas are (pre-)shaped also on regional and local level. In the current programming period 2012–2020, the cohesion policy (incorporating the European Regional Development Fund, European Social Fund and Cohesion Fund) has also been streamlined according to the Europe 2020 development goals (e.g. Garretsen et al. 2013; McCann 2015; McCann and Ortega-Argilés 2013). Without digging too deep into the history and assessment of this approach (e.g. Avdikos and Chardas 2016; Marques and Morgan 2018; Mendez 2013), it is safe to say that the alignment towards the EU2020 goals led to a narrowing of thematic choices for the programmes, especially because of the strict monitoring and goal fulfilment procedures. The EU2020 goals were broken down into 11 thematic fields (Fig. 4.2), of which not all were open to all sub-programmes[1].

According to these overall themes, national cooperation agreements were developed, assessing national needs and action fields in relation to the overall topics. The operational programmes on national/regional level were then drafted following these cooperation agreements, underlying the targets named in the national documents. These documents are the nodal points between the Commission's targets and the regional and local needs.

Given this approach from the top level of the policymaking hierarchy, regional and local initiatives utilising specific potentials (as well as addressing specific problems) are thereby limited, as the agenda-setting process has been structured top-down

[1] Investment from the ERDF will support all 11 objectives, but 1–4 are the main priorities for investment; main priorities for the ESF are 8–11, though the Fund also supports 1–4; the Cohesion Fund supports objectives 4–7 and 11 (http://ec.europa.eu/regional_policy/en/policy/how/priorities).

and the thematic focus has been narrowed in advance. For instance, the topic of demographic change cannot be addressed directly by the stipulated priorities. Nevertheless, when drafting the operational programmes there is a wide consultation process involving various actors from different administrational levels, where regional and local priorities can be taken into account when drawing up the framing documents—thereby re-opening and supplementing the agenda-making process on a lower governance level according to the prerequisites. However, this assumes that regional and local actors are already aware of their endogenous potentials. Accordingly, it is of special interest to consider how the regional and local policy levels act on the identification and utilisation of endogenous potentials. In this context we refer in the following analysis to Hooghe and Marks (2001), who characterise different forms of governance arrangements in regional development contexts. They differentiate between approaches of strong state involvement through specific interventions and cases with no clear, overriding responsibility development where self-organisation is the predominant form of governance (e.g. Chhotray and Stoker 2009).

In the context of the main question of this chapter—utilisation of endogenous potentials on the local and regional level—three sub-questions need to be answered: What role do endogenous potentials play in the regions? Which governance arrangements are crucial for agenda setting when utilising place-based potentials? And finally, which kinds of implementation mechanisms were created? The analysis of the case studies will highlight some important points for a more general understanding of the context. The description cannot extensively cover all the framework conditions in the region, nor all aspects of potentials possible.

4.4 Case Studies

In the following section, the analysis will introduce and discuss two case studies from (post-) industrial regions in non-metropolitan settings in Central Europe. Both regions have been extensively researched by the authors in the context of EU-funded projects over the last decade, in which the authors organised workshops and performed project tasks, and which have led to a deeper insight into issues of regional development in these places.

The focus in some of these collaborations has been on governance arrangements and the utilisation of endogenous potentials (Harfst 2015; Wirth et al. 2012). The authors have worked together with regional partners in various projects, whereby the research agenda has loosely followed an 'action research approach' (Greenwood and Levin 2007), incorporating the research partners in practical tasks with regional actors, but also enabling them to reflect on the overall aims of the project. The methodology of the approach has been described in some detail elsewhere (Harfst and Osebik 2015). Here, the research has been backed up by document research, taking into account regional policy papers and project outputs. Additionally, interviews with regional actors have been performed at various stages of the project work. The analysis will—in accordance with the research question—focus on governance arrangements in both case study regions, mainly leaving out a wider discussion of framework conditions in both the regions.

The overall focus of this chapter is partly based on research results from the ERDF-funded INTERREG-B project InduCult2.0 (INTERREG CE, 2016–2019, InduCult 2017) on the utilisation of industrial culture for regional development in (post-) industrialised regions of Central Europe.

Industrial Heritage as an Endogenous Potential in the County of Zwickau

The Landkreis (county) of Zwickau is situated in the state of Saxony in the eastern part of the Federal Republic of Germany. It is geographically located between the Saxon Uplands (Sächsisches Hügelland) to the north and the 'Ore Mountains' (Erzgebirge) to the south. In its current form, the Landkreis is an amalgamation of three smaller units brought together in the framework of an administrative reform in 2008. It covers an area of 950 km^2 and had 322,000 inhabitants in 2016. The city of Zwickau is the largest municipality of the Landkreis with a population of 90,500 in the same year. It is the region's administrative, industrial and cultural centre. Nevertheless, with Crimmitschau, Glauchau and Werdau there are other cities with populations between 19,000 and 23,000 inhabitants (2016). The extremely high population density of 339 inh./km^2 (all demographic data: DM Sachsen 2017a) is a hint of the old industrial character of the area, which is mainly based on the textile industry, coal mining, mechanical engineering, and for more than 100 years also the automotive industry (Karlsch and Schäfer 2006).

In recent history, the Landkreis has been challenged by profound structural changes. Whereas coal mining ended in the 1970s, the most intensive shock hit the region after German unification. The shrinkage of the textile industry was dramatic, and mechanical engineering and car production also declined substantially. But unification opened up new perspectives as well: today Volkswagen Saxony GmbH in Zwickau is the biggest enterprise in western Saxony with about 10,000 employees (VW Sachsen 2017b).

Against this economic background it is apparent that the Landkreis has a considerable industrial heritage. The relics are manifold, including impressive examples of industrial architecture, museums, archives, and also mining landscapes with huge waste dumps, subsidence areas and old shaft frames. In the 1990s—the years after the unification with industrial decline producing high unemployment—this heritage did not play a progressive role in political concepts on the regional and local level. In fact, the tangible remains of the old industries were often seen as brownfields and environmental risks. From the end of the 1990s, this perspective began to change step by step. One of the first European projects in the region—Revitalisation of Towns in former Coal Mining Areas (1998–2001; INTERREG-C programme)—dealt with the management of underground stabilisation, depression areas inside the city, and the Martin Hoop Shaft area in Zwickau characterised by 'wasteland and scattered small industries'(REVI 2001: 47). Back then Zwickau used national funding to rehabilitate a former coking plant area and to erect a shopping centre.

An important step towards bringing the industrial heritage onto the political agenda of the region involved the Wirtschaftsregion (Economic Region) Chemnitz-Zwickau GmbH (WiReg), a public management unit founded in 1995 by the cities of Chemnitz and Zwickau and three surrounding counties to support the economic transition of the whole agglomeration area. The development concept of the WiReg—which won a national competition by the German Building Agency in 2000 (Adam 2001)—highlighted the industrial heritage of the region as an integral part of regional development for the first time. From today's perspective, this can be seen as the starting point of a debate about industrial history, heritage and culture as a development factor of the region.

Subsequently, several initiatives of the Landkreis and City of Zwickau aimed to develop its industrial heritage in the framework of European development programmes. From 2003 to 2006, the City of Zwickau joined the INTERREG III-B project READY about rehabilitation and development in former mining regions (Wirth and Lintz 2007). The core element of the project was a feasibility study to prove the suitability of the winding tower at Martin Hoop Shaft for a leisure and sports centre. Later on, the city of Zwickau took part in the INTERREG IV-B project ReSource (2009–2012), dealing with post-industrial strategy development (Harfst and Wirth 2012). For the first time, the Landkreis Zwickau now took the lead in a European project. It established a new position in the regional administration, called Management of European Projects, making it a part of the economic promotion department.

In a subsequent step, the Landkreis initiated the INTERREG IV-B project SHIFT-X—valorising industrial culture. Industrial heritage is characterised as a promoter of economic and social transition in the region and as a pre-condition of sustainable and endogenous development in structurally disadvantaged regions (Shift-X 2014). At present, the latest initiative of the Landkreis is the INTERREG V-B project InduCult2.0 (InduCult 2017). This project aims less at the implementation of individual projects but rather at generating regional strategies on industrial culture in Zwickau and the other European regions involved. The project tries to tackle one problem of the Zwickau region: the weak involvement of civil society actors in the promotion of industrial heritage. Consequently, the InduCult project intends to connect industrial heritage with the modern economy and tries to involve actors from different fields of activity. Recently, a so-called focus group, including 25 public bodies on the local and regional level, museums, small- and medium-sized companies, higher education institutes, and creative workshops, has launched a corresponding initiative (Landkreis Zwickau 2017).

Recent activities of regional actors have also been directed towards a major upcoming event: the Federal Exhibition of the Free State of Saxony in 2020 under the headline 'Man – Industry – Culture'. Although the 6-month-long event follows a decentralised strategy with several destinations in south-west Saxony, Zwickau is designated to become the location of the central exhibition. It shall present one of the heartlands of industrialisation in Germany and connect this with future perspectives. The Zwickau region not only hopes to become a Mecca of industrial heritage for 6 months, but also to present itself as a modern, innovative and smart place.

Utilisations of Endogenous Potentials in the 'Steirische Eisenstrasse'

The 'Steirische Eisenstrasse' (Styrian Iron Route) is an old-industrial region in the province of Styria (Austria). The region was shaped over centuries by mining and steel production. From the nineteenth century onwards it was one of the industrial centres of Austria. It is still the location of Europe's biggest operating ore mine in the town of Eisenerz at the Erzberg (Fig. 4.3), whose silhouette has become a national symbol (Posch et al. 2004) and is now also exploited for tourism. Today the region is home to the VOEST Alpine steelmaker with its rolling mill in Leoben-Donawitz. Nevertheless, due to globalisation and modernisation processes industrial production has steadily lost importance for the region—especially in terms of employment—over the past 60 years. The economic transformation process has brought about a variety of social changes, most notably profound demographic change in the region, and here especially the outmigration of younger people (Zimmermann and Janschitz 2004).

Against this background the region has tried to re-invent itself over the past 20 years as a more tourism-friendly region, based on the natural landscape, as well as promoting the industrial heritage of the region. In the case of the 'Steirische Eisenstrasse', the industrial and mining heritage was identified by relevant stakeholders as an asset early on, with mainly tangible elements being identified as having potential—especially for altering the image of a 'rusty and dirty' place (see Fig. 4.3). A flagship project in the past was the 'Museumsverbund Steirische Eisenstrasse', which unites 13—previously independent—museums in the region in one association (Eisenstrassenmuseen 2017). For more than 10 years

Fig. 4.3 The Erzberg in Eisenerz and its characteristic extraction levels (Picture: Simić)

now, the museums have cooperated in the fields of fundraising, marketing and human resources. The existence of this cooperation, which came about with the (financial) backing of state and regional politicians, shows the support that cultural heritage has in the region. The association is an important structure which contributes to the maintenance of each of the museums involved, although they still retain their independence. Nevertheless, even in this region, especially with the negative demographic changes in progress and low investment incentives, many buildings have been abandoned and there are difficulties to find appropriate after-uses. The steering of the process has also proven challenging, despite consensus among decision-makers and a long-term strategy in place. One example of failure is the state of the 'Erzbergbahn', a local museum railway line set in an impressive mountain panorama with high touristic potential. Here a lack of funding, deteriorating infrastructure and problematic property issues involving the national train operator have basically led to the abandonment of the project (Fischer 2014).

The Association Steirische Eisenstrasse e.V. is the main driver for the overall development strategy, organising and leading the process. It is a non-profit organisation founded by 15 municipalities, which coordinates and facilitates a variety of activities connected to industrial heritage, tourism and fundraising (Eisenstrasse 2017). In contrast to the case study above and its organisational structures, the Steirische Eisenstrasse does not follow the administrative units of Styria but is an association of municipalities from two political districts, building on the idea of a region with a common past and common understanding.

This approach was fostered by the presence of some surviving large-scale industries and the mainly positive local attitudes towards the industrial past. Moreover, both the Montan-Universität Leoben—with its industry-related focus—and the University of Graz—via its regional development competence—have fostered institutional capacities. This approach has been strategically enshrined in the consecutive regional 'LEADER' strategies[2]. Via national and European funding instruments, the region has been able to conserve and promote a variety of its rich industrial heritage in various ways (museums, events and festivities, etc.) (Osebik 2012). In the latest LEADER strategy ('High Tech versus High Feeling', 2014–2020), the high feeling is to be understood as an attitude towards life and a prerequisite for economic, high technology-based initiatives. Here also a reinforced pioneer spirit is seen as pivotal for future developments (LAG Steirische Eisenstraße 2015). Thus, in recent years a notable shift towards participation-based citizen involvement strategies can be noticed.

[2]The LEADER programme ('Liaison Entre Actions de Développement de l'ÉconomieRurale') is an European Union initiative to support rural development projects initiated at the local level since 1991.

4.5 Discussion

The case studies in this chapter underline the importance that endogenous potentials can assume in regions that face deep processes of structural changes and lack major development impulses from outside. In both regions, the utilisation of tangible and intangible industrial remains has taken a prominent place in the strategic development initiatives. The utilisations range from very different approaches incorporating former industrial buildings, sites and other physical remains to approaches that build on specific traditions, skills and mind-sets. Over time the initiatives have involved a growing range of network actors, including still-existent industries, the creative sector, schools and museums, thereby enlarging the traditionally narrow actor networks in the regions considerably. In both examples, utilising the specific place-based potentials of the regions, the actors were able to generate extra attention via the topic of industrial culture on state and national level, in the Zwickau example via the upcoming state exhibition (Landesausstellung) and in the case of the Steirische Eisenstrasse via the Erzberg. In both cases, there has also been a noteworthy shift from site-based interventions towards a more strategic focus in the implemented measures.

Having highlighted these similarities, the cases also show very different governance arrangements in utilising their place-based potentials. In the case of Zwickau, the timeline shows that the district office has slowly but steadily built up capacities, enabling it to lead European projects in connection to the use of the region's industrial remains. By creating fixed positions within its administration for this purpose, the district has firmly assumed overall responsibility for the agenda setting and the ways in which potentials are framed and used in the region, and is set to continue to do so in the future. In the Austrian case, this centralised management function is missing. Here the impetus of utilising territorial potentials lies with the regional LEADER action group, and is thereby set firmly on the lower tiers of policymaking, without a major policymaking apparatus in the background. Also for LEADER a wider range of local actors are involved, but much more in a non-formal, project-based setting. Despite this rather low policymaking capacity of the group, the region also shows long continuity regarding the topic of industrial culture, creating a stable strategy-based initiative levering EU funds despite the low involvement of the higher policy levels. Here the persistence of actors and a stable actor network seem to be the key for a continued use of place-based potentials. This can also be interpreted as a strength of the initiative, whereas the Zwickau region is still struggling for stronger societal involvement.

Looking at the examples regarding the implementation mechanisms created and utilised, it becomes clear that both regions rely mostly on project-based funds from outside when pursuing place-based potentials, thereby facing a fragile funding situation, which was seemingly mediated by the capacities and networks created—and resulted in follow-up projects. Interestingly in both examples there is a change in the policy agenda—from single, direct interventions at certain sites, to creating broader, post-industrial development perspectives via strategy development and networking, thereby focusing more on policymaking. This also corresponds with a gradual opening of actor networks involved in the projects and agenda setting.

While both forms of governance—in the one case managed by a public body, in the other case by a non-profit network—have yielded remarkable results, it is clear that each mode has distinct weaknesses. The Austrian example is clearly dependent on the networking capacities of certain key actors whose withdrawal from the network would be difficult to compensate. With concepts not being entrenched in official policy documents on state-level and the tight-knit LEADER network, the utilisation of post-mining potentials sometimes remains isolated from other initiatives in the state of Steiermark.

In the case of Zwickau, the integration of approaches to exploiting post-mining potentials into the more formal district structures has led to weaknesses related to including certain groups in the process (i.e. civil society, business). Strategic aims are accordingly not pursued by a broad base of actors, implying a danger of 'lock-in' and a lack of innovation.

Despite these shortcomings, both examples show that cooperation influences capacity building in the region in various ways: (1) a better perception of regional requirements by state authorities, (2) purposeful access to state and European funding and (3) an opportunity for larger joint projects and international collaboration, knowledge and experience exchange. As the case studies show, such cooperation on the utilisation of post-mining potentials has significantly increased regional capacities to act, whether with or without specific strategic development concepts being in place. Valuable learning effects and gains in knowledge can be noted, mainly through the inclusion of other actors and the opportunity to employ extra staff continuously through project funding. This has led to the successful initiation of new projects and the inclusion of new network partners.

4.6 Conclusion

The cases presented in this chapter have highlighted current development approaches of non-agglomeration, post-industrial regions in Central Europe, focusing on the identification and utilisation of place-based potentials. An insight was given into the agenda setting regarding endogenous potentials at the top level of policymaking, and some lessons learned from the locally led, place-based development initiatives were discussed. The chapter has shown that the thematic focus when dealing with endogenous potentials has been already pre-shaped in the European Union cohesion policy programme. This top-down shaping via various policy frameworks is mediated in local and regional governance settings by adapting and translating these programmes into practical, regional and local actions.

On the regional level, we can identify—in accordance with Hooghe and Marks (2001)—two different policy models in the case study regions, one firmly set in the regional administrative governance machine on the district level, while the other is built around a network of actors in connection to LEADER. Both examples show a strong focus on the utilisation of (post-) industrial place-based potentials for regional development. In both cases the identification and utilisation has matured over time, first, in regard to the kind of policy agenda pursued (from individual projects to strategic policy interventions) and, second, in the way both regions have built important capacities for handling European projects and acquiring additional funding, also enlarging their regional actor networks considerably (policy implementation). In this context, both regions were able to use mainly European funding to develop and shape a new development strategy via the utilisation of place-based (post-) industrial potentials. The findings thereby underline the importance of European cohesion funding as such for practical development on a regional level, via capacity building and practical utilisations. Nevertheless, both examples are strongly reliant on project funding from cohesion fund sources. It remains to be seen how this development approach continues when outside funding is missing or—especially valid for the Austrian case—important network actors drop out. Overall the text highlights the importance place-based potentials can assume, especially in regions where there is a lack of major outside impulses for growth.

The here described findings have multiple interrelations to the thematic focus of this volume. Firstly, it becomes obvious that a strong strategic approach towards the use of the industrial heritage seems hardly possible without the organisational and conceptual capacity of the public sector. Secondly, the chapter supports the hypothesis that without the support of EU funds a powerful valorisation of heritage aspects on a regional and local level would be illusory. Thirdly, in multi-level governance systems, capacity building often happens by shaping policy frameworks on European and national levels, as well as translating and implementing these preconditions into the regional and local context. Last but not least, on the regional and local level specific governance settings are of crucial relevance, formed by actors, their networks, capacities and policy traditions.

References

Adam B (2001) Regionen der Zukunft. Spuren eines Wettbewerbs. http://www.bbr.bund.de/BBSR/DE/FP/MORO/Initiativen/2004undFrueher/WettbewerbRegionenZukunft/DL_Adam.pdf?__blob=publicationFile&v=3

Amin A, Thrift N (1995) Institutional issues for the European regions: from markets and plans to socioeconomics and powers of association. Econ Soc 24:41–66

Avdikos V, Chardas A (2016) European Union Cohesion Policy Post 2014: more (place-based and conditional) growth—less redistribution and cohesion. Territory Polit Gov 4:97–117

Barca F (2009) An Agenda for a reformed cohesion policy. A place-based approach to meeting European Union challenges and expectations

Chhotray V, Stoker G (2009) Governance theory and practice—a cross-disciplinary approach. Palgrave Macmillan

Damsgaard O, Lindquvist M, Roto J, Sterling J (2009) Territorial potentials in the European Union. Nordregio Working Paper 2009. Stockholm, p 6

Danielzyk R (2012) Der raumordnungspolitische Metropolendiskurs – Konstruktion von (neuen) Peripherien? disP—Plan Rev 48:27–33

Dunford M, Smith A (2000) Catching up or falling behind? Economic performance and regional trajectories in the "New Europe". Econ Geogr 76:169–195

Eisenstrasse (2017) Association's homepage. http://www.eisenstrasse.co.at/. Accessed 13.12.2017

Eisenstrassenmuseen (2017) Association's homepage. http://www.eisenstrassenmuseen.at/. Accessed 15.12.2017

EU (2011) Territorial Agenda of the European Union 2020. Towards an inclusive, smart and sustainable Europe of diverse regions [WWW Document]. http://www.nweurope.eu/media/1216/territorial_agenda_2020.pdf. Accessed 6.23.17

European Commission (2010) EUROPE 2020. A European strategy for smart, sustainable and inclusive growth [WWW Document]. http://ec.europa.eu/eu2020/pdf/COMPLET%20EN%20BARROSO%20%20%20007%20-%20Europe%202020%20-%20EN%20version.pdf. Accessed 9.1.2018

Fischer W (2014) The mining and industrial heritage presented by the Association of Museums of the Styrian Iron Road. A story of success? In: Rust, regeneration and romance: iron and steel landscapes and cultures, conference proceedings. International Institute for Cultural Heritage, Ironbridge, 10–14 July 2013. Ironbridge, Telford

Garretsen H, McCann P, Martin R, Tyler P (2013) The future of regional policy. Camb J Reg Econ Soc 6:179–186

Greenwood D, Levin M (2007) Introduction to action research: social research for social change, 2nd edn. Sage, Thousand Oaks

Gruenewald DA, Smith GA (2014) Place-based education in the global age: local diversity. Routledge, New York

Harfst J (2015) Utilizing the past: valorizing post-mining potential in Central Europe. Extr Ind Soc 2:217–224

Harfst J, Osebik D (2015) Social learning in transnational projects—lessons from European territorial cooperation projects. https://www.researchgate.net/publication/281223383_Social_learning_in_transnational_projects_-_lessons_from_European_territorial_cooperation_projects

Harfst J, Simić D (2017) Industrial Culture as an emerging topic in regional development? pp 145–152

Harfst J, Wirth P (2011) Structural change in former mining regions: problems, potentials and capacities in multi-level-governance systems. Procedia Soc Behav Sci 14:167–176

Harfst J, Wirth P (2012) Zwickau-Lugau-Oelsnitz (Germany)—the long shadow of mining. In: Wirth P, Černič Mali B, Fischer W (eds) Post-mining regions in Central Europe. Problems, potentials, possibilities. oekom, Munich, pp 118–128

Harfst J, Wirth P (2014) Zur Bedeutung endogener Potenziale in klein- und mittelstädtisch geprägten Regionen—Überlegungen vor dem Hintergrund der Territorialen Agenda 2020. Raumforsch Raumordn 72:463–475

Hooghe L, Marks G (2001) Multi-level governance and European integration. Rowman & Littlefield, Oxford

InduCult (2017) Project homepage. http://interreg-central.eu. Accessed 14.12.2017

Jeffery C (2015) The regional dimension of the European Union: towards a third level in Europe? Routledge, London, New York

Karlsch R, Schäfer M (2006) Wirtschaftsgeschichte Sachsens. Edition Leipzig

LAG Steirische Eisenstraße (2015) Die Lokale Entwicklungsstrategie der LEADER-Region Steirische Eisenstraße 2014-2020: High Tech versus High Feeling [WWW Document]. Steirische Eisenstrasse. http://www.steirische-eisenstrasse.at/lokale-entwicklungsstrategie/eu-periode-2014-2020/. Accessed 24.1.2018

Landkreis Zwickau (2017) Authentisch und dynamisch. Argumente für eine lebendige Industriekultur im Landkreis Zwickau [WWW Document]. Industrikultur in Sachsen. http://www.industriekultur-in-sachsen.de/fileadmin/user_upload/11_Argumente_f%C3%BCr_eine_lebendige_Industriekultur_im_Landkreis_Zwickau.pdf. Accessed 10.1.2017

Lazzeroni M, Bellini N, Cortesi G, Loffredo A (2013) The territorial approach to cultural economy: new opportunities for the development of small towns. Eur Plan Stud 21:452–472

Maier G, Tödtling F, Trippl M (2006) Regional- und Stadtökonomik/Gunther Maier; Franz Tödtling; 2: Regionalentwicklung und Regionalpolitik, 3., aktualisierte und erw. Aufl. ed, Springers Kurzlehrbücher der Wirtschaftswissenschaften Springer, Wien [u.a.]

Marques P, Morgan K (2018) The heroic assumptions of smart specialisation: a sympathetic critique of regional innovation policy. In: Isaksen A, Martin R, Trippl M (eds) New avenues for regional innovation systems—theoretical advances, empirical cases and policy lessons. Springer, New York

Martin R, Sunley P (1998) Slow convergence? The new endogenous growth theory and regional development. Econ Geogr 74:201–227

McCann P (2015) The regional and urban policy of the European Union: cohesion, results-orientation and smart specialisation. Edward Elgar Publishing

McCann P, Ortega-Argilés R (2013) Modern regional innovation policy. Camb J Reg Econ Soc 6:187–216

Mendez C (2013) The post-2013 reform of EU cohesion policy and the place-based narrative. J Eur Publ Policy 20:639–659

Moulaert F, Sekia F (2003) Territorial innovation models: a critical survey. Reg Stud 37:289–302

Myrdal G (1957) Economic theory and underdeveloped regions. Duckworth, London

Oevermann H, Mieg HA (eds) (2014) Industrial heritage sites in transformation: clash of discourses, 1st edn. Routledge, New York

Osebik D (2012) Steirische Eisenstraße (Austria)—the region surrounding an outstanding mining landmark. In: Wirth P, Černič Mali B, Fischer W (eds) Post-mining regions in Central Europe. Problems, potentials, possibilities. oekom, Munich, pp 79–91

Perlik M (2001) Alpenstädte - Zwischen Metropolisation und neuer Eigenständigkeit. Arbeitsgemeinsch. Geographica Bernensia, Bern

Petrakos G, Rodríguez-Pose A, Rovolis A (2005) Growth, integration, and regional disparities in the European Union. Environ Plan A 37:1837–1855

Pinder D (2017) Regional economic development and policy: theory and practice in the European Community. Routledge

Pollack MA (1997) Delegation, agency, and agenda setting in the European Community. Int Organ 51:99–134

Posch A, Steiner G, Risopoulos F (2004) Die Erzherzog-Johann-Fallstudie—ein inter- und transdisziplinäres Lehr- und Forschungsprojekt der Kulturlandschaftsforschung. Österreichisches Bundesministerium für Bildung. Wien

Princen S (2007) Agenda-setting in the European Union: a theoretical exploration and agenda for research. J Eur Publ Policy 14:21–38

REVI (2001) Revitalisation of towns in former coal mining areas (REVI). Stadtentwicklung Südwest GmbH (STEG), Dresden

Romer PM (1994) The origins of endogenous growth. J Econ Perspect 8:3–22

Sachsen DM (2017a) Demografiemonitor Sachsen [WWW Document]. http://www.demografie.sachsen.de/monitor/html/atlas.html. Accessed 1.10.18

Sachsen VW (2017b) Volkswagen Sachsen GmbH [WWW Document]. Volkswagen Sachsen GmbH. http://www.volkswagen-sachsen.de/de/unternehmen/zahlen-und-fakten.html. Accessed 1.10.18

Shift-X (2014) Project homepage. http://www.shiftx.eu/. Accessed 21.12.2017

Stough RR, Stimson RJ, Nijkamp P (2011) An endogenous perspective on regional development and growth. In: Kourtit K, Nijkamp P, Stough RR (eds) Drivers of innovation, entrepreneurship and regional dynamics. Springer, Berlin, Heidelberg, pp 3–20

Wirth P, Lintz G (2007) Strategies of rehabilitation and development in European mining regions. In: Good (best) practice cases in regional development after mining and industry, Grazer Schriften der Geographie und Raumforschung. Universität Graz, pp 75–85

Wirth P, Černič Mali B, Fischer W (eds) (2012) Post-mining regions in Central Europe. Problems, potentials, possibilities. oekom, Munich

Wirth P, Elis V, Müller B, Yamamoto K (2016) Peripheralisation of small towns in Germany and Japan—dealing with economic decline and population loss. J Rural Stud 47:62–75

Zimmermann FM, Janschitz S (2004) Erfolgs- und Misserfolgsfaktoren bei der Umstrukturierung von traditionellen Bergbaugebieten. Das Beispiel Eisenerz/Österreich. In: Sächsisches Staatsministerium des Innern (ed) Neue Landschaften, Bergbauregionen im Wandel. Dokumentation des Fachkongresses auf der Euregia 2004 in Leipzig. Dresden, pp 30–41

Chapter 5
Energy Sensitive Spatial Planning as a Public Sector Tool Towards Sustainable Economic and Territorial Development

Elena Dimitrova, Milena Tasheva-Petrova, Angel Burov, and Irina Mutafchiiska

5.1 Introduction

Spatial planning of the twentieth century developed as a public policy instrument for regulating societal relations in territorial development. The established political system in Bulgaria and the centrally planned economy in the period 1948–1989 relied on highly centralized "top-down" planning system claiming the ultimate priority of public interest. The political changes in CEE after 1990s brought about challenges of redefining the priorities and responsibilities of the public sector; they brought about the need for decentralizing Bulgarian planning system and introducing a more democratic and participatory planning process. The changes made in the planning legislation of the country and the adopted strategic policy documents were aimed at empowering the local level of governance. The global and European process towards sustainable development also confirmed the importance of local governance and spatial planning for holistically addressing the social, economic, and environmental aspects of development in Agenda 21 (UN 1992). The energy topic was addressed by the sustainability debate as early as the Rio Summit and was also clearly traceable as a crucial driving factor with an impact on all the dimensions of sustainable development. The current chapter aims at outlining the linkages between two topics—spatial planning and energy efficiency, on the path towards sustainability, trodden by the local level of public governance in Bulgaria during the first decade after the country's accession to the European Union in 2000.

E. Dimitrova (✉) · M. Tasheva-Petrova
Department of Urban Planning, University of Architecture, Civil Engineering and Geodesy, Sofia, Bulgaria

A. Burov · I. Mutafchiiska
Association of Bulgarian Urban and Regional Planners (ABURP), Sofia, Bulgaria

© Springer International Publishing AG, part of Springer Nature 2019

M. Finka et al. (eds.), *The Role of Public Sector in Local Economic and Territorial Development*, EAI/Springer Innovations in Communication and Computing, https://doi.org/10.1007/978-3-319-93575-1_5

5.2 Meeting Points of "Urban" and "Energy" on the Path to Sustainable Development: The Global, European, and Bulgarian Context

The "urban" and "energy" topics have both been increasingly acknowledged as key ones in the global striving for sustainable development (SD). In parallel, spatial planning and energy transition[1] processes face similar operationalization challenges and require further research and policy efforts for practical implementation, monitoring, and evaluation.

The Global Political Framework

Energy and spatial planning appeared as interlinked topics in the political documents on sustainable development as early as the Rio Summit in 1992. There was a clear message to architects and planners about their responsibilities in the process towards a more sustainable future in Agenda 21, which claimed that "shelter for all" should be provided in a world of one billion homeless people; spatial planning and management should be improved and integrated approaches to building the urban infrastructure (electricity, water supply and sewage, waste collection, etc.) should be developed. The importance of planning and design results at various scales—from the urban system to the single building, was discussed alongside the actors involved and their interactions during the planning process itself. The messages were further detailed in numerous more recent global policy documents such as the Sustainable Development Goals (UN 2015) and the New Urban Agenda (UN-HABITAT 2016).

The European Process: Policy and Action

Global messages have been further interpreted and implanted into the European cultural context through numerous international and EU documents. The urban dimensions of sustainability have been discussed by the Aalborg Charter on Sustainable Cities (ICLEI 1995); the European Spatial Development Perspective (ESDP 1999); the Goteborg Sustainable Development Strategy of EU (EC/COM 2001, 2005); the Leipzig Charter on Sustainable European Cities (EC 2007); the

[1] Defined as a long-term structural change in energy systems aimed at developing a secure, affordable, and climate-friendly energy system better able to manage and balance dynamic patterns of supply and demand from a national and even European level right down to a local level. This must enhance both national and European energy security while ensuring energy equality: affordable energy accessible by all citizens (Partnership on Energy Transition 2017).

Territorial Agenda of the European Union 2020 (EC 2011); the Urban Agenda for the EU (EC 2016b), etc. The two topics were in most of them closely linked: The Leipzig Charter was the first to explicitly stress the need for improving the energy efficiency of cities, and especially on the key importance of the compact urban structures for urban sustainability. It outlined four priority issues in the SD process—political debate on implementing the principles at all levels, the introduction of integrated planning and balanced territorial organization; the need for holistic strategies and coordinated action at all levels, the modernization of infrastructure networks and improvement of the energy efficiency of the cities. It recommended four key measures for increasing energy efficiency at the urban level—improvement of public transport, refurbishment of residential buildings, efficient resource use; reasonable use of the territory resulting in compact settlement structures. Spatial planning was there envisaged as an activity that "prevents urban sprawl by strong control of land supply and of speculative development." Toledo Declaration (EC 2010) argued that "*the core battle for urban sustainability lies in achieving the maximum possible eco-efficiency in the existing urban fabrics of cities*." The outlined key issues there comprise: (a) reducing transport needs by the promotion of proximity and mixed-uses schemes, while at the same time stimulating a more sustainable mobility; (b) improving the management of energy and material resources and flows in the city; (c) recycling land by means of urban regeneration, the redevelopment or reuse of abandoned, derelict, or unused areas, etc. It also stressed that "*strategies and actions should be unitary, and address the whole complexity of urban development, taking into account the role of each part of the city in the whole structure*."

In parallel, the energy-focused debate identified the social and spatial dimensions of the envisaged energy transition. The Energy Roadmap 2050 (EC/COM 2011) acknowledged that "*not only the energy system but society as a whole need to be dramatically more energy efficient*"; it also claimed that "*the path towards a new energy system has a social dimension that requires encouraging social dialogue.*" Stating that the role of local organizations and cities will be much greater in the energy systems of the future, the Energy Roadmap 2050 (EC/COM 2011) foresees the need to answer "*questions on to what extent urban and spatial planning can contribute to saving energy in the medium and long term*." The updated European Strategic Energy Technology Plan (SET Plan) (EC 2015) aimed at accelerating the development and deployment of low-carbon technologies, acknowledges that "*It is essential that socio-economic impacts of the energy system transformation and the factors that determine people's interaction with the energy system are better understood*." An overview of EU policies in the energy efficiency field in "Good Practice in Energy Efficiency: For a sustainable, safer and more competitive Europe" (EC 2016a), among the lessons learned explicitly mentions that "*consumer engagement and acceptance is key to delivering energy savings by behavioral change*" and recommends to policy-makers "*to effectively target, monitor and evaluate their measures and actions*" as particularly useful. While discussing the aspects of setting the right public policy framework, the authors acknowledge that measurable targets are key ones to guiding and coordinating any policy interventions. It is also acknowledged there that many regional and local authorities have adopted ambitious energy efficiency and sustainable energy strategies. Moreover, this has been achieved in many instances, through voluntary commitments in the context of EC initiative Covenant of Mayors.[2] The Directive on Energy efficiency in its Article 17 (EC 2012) stated that *"public bodies at national, regional and local level should fulfil an exemplary role as regards energy efficiency."* It also acknowledged that a number of municipalities and other public bodies in the Member States had already applied integrated approaches to energy saving and energy supply. The sustainable energy action plans developed under the Covenant of Mayors initiative and the integrated urban approaches, going beyond individual interventions in buildings or transport modes were explicitly estimated as very successful. The Directive recommended that Member States encourage municipalities and other public bodies to adopt integrated sustainable energy efficiency plans with clear objectives; to involve citizens in their development and implementation; to adequately inform them about the progress in achieving objectives.

Urban structures and urban planning has been already discussed as important sustainability factors in planning literature (Campbell 1996; Naess 2006); research reports have repeatedly focused on urban sprawl (EEA 2006; EEA-FOEN 2016); European studies have already identified mobility and transport issues and their linkages to urban development (Institute for Energy and Transport 2012), outlining the importance of urbanization modes and transport to the efficient use of land and energy resources. Claiming the importance of place-based policy interventions, Davoudi argues in 2010 that on the ground of efficiency, place matters because market failures and the unintended consequences of government policy impact different places in different ways. Development aspects in peripheral areas in SEE were also claimed to require higher sensitivity to social equity and territorial cohesion (Dimitrova and Burov 2010). Efficiency is also a research issue still to focus upon more carefully—broader considerations than the economic ones have been recommended in the evaluation of planning concepts (Miller and Patassini 2007). The increasing interest of social sciences to energy transition processes has provided motivation for undertaking interdisciplinary research in the field (Biresselioglu et al. 2017; EEA 2013). A growing body of academic research aims at understanding and bridging the gaps between theory and practice in the field (e.g., Healey 1997; EEA 2001; Khakee 2003; Girard et al. 2007; Dimitrova 2007; POINT project 2007). Yet, the implications of energy efficiency to the regional and local spatial planning levels are only generally covered by research projects. When presenting the case study of Bilbao and explicitly discussing urban transformations *"in relation to barriers and opportunities for low-carbon transitions,"* Olazabal and Pascual (2015) note that the planning of the city has not yet included energy criteria. The authors also acknowledge that *"although being necessary, low-carbon transition strategies are only part of a broader set of solutions for achieving sustainable urban development."* Furthermore, the creation of an enabling environment including

[2] Launched in 2008 (and transformed into the Covenant of Mayors for Climate and Energy in 2015, to endorse and support the efforts made by local authorities in the implementation of sustainable energy and climate policies (http://www.covenantofmayors.eu/).

education and awareness-raising, and the systematic development of integrative policies with broader sectors (agriculture, transportation, water management, and urban planning) are recommended in order to tailor and implement policies that extend beyond the research and development and support technology deployment (Sen and Ganguly 2016).

Addressing the ambitious aim of creating smart, integrated energy systems effectively working at all levels, the Partnership on Energy Transition (2017) regards energy as *"binder and enabler of other partnerships, most of them with explicit urban dimensions, especially: housing; air quality; climate adaptation; urban mobility; urban poverty; circular economy; sustainable land use."* The Partnership, in its Orientation Paper on Energy Transition, outlines the demand *"to ensure sustainable urban development plans and approaches, with energy masterplanning playing a key role."* It also outlines the major urban challenges in need for inter- and transdisciplinary approaches: *"the complex ownership structure of many types of buildings, both residential and non-residential, and the lack of capacity for cost-effective and efficient energy storage solutions, both at the building and system level."* Building an understanding about the life cycle costs of energy transition—construction, management, and operation of the urban system is claimed an important opportunity to take.

Monitoring and measuring of energy efficiency has been increasingly focused upon in line with the growing understanding about their importance to the practical implementation of the sustainability concept and the need for new instruments to realize it. The REViSITE Coordination Action (2012) has developed a cross-sectoral roadmap, outlining a common vision, strategic research agenda, and implementation action plan for multi-disciplinary, ICT-enabled energy efficiency. Yet, the *Efficiency First* Foundation (2016) argues that *"the Commission's current methodology to assess the impact of different energy efficiency measures does not do energy efficiency justice as it looks at the cost of energy efficiency targets from the perspective of individual consumers, rather than analyzing the costs and benefits to society as a whole, or taking into account the role that policy can play in overcoming barriers to realizing these benefits."*

The National Context of Bulgaria

Two characteristics of Bulgarian urban structures could be considered important in influencing their energy consumption—their spatial identity and prevailing mobility patterns. Up to mid twentieth century most Bulgarian cities followed the traditional European spatial pattern—a radial structure with a lively multifunctional center keeping its attractiveness for all the inhabitants and visitors due to the cultural, administrative, and shopping premises there, and complemented by an efficient radial structure of the public transport providing an easy access to the city center (Bertaud 2004). The socialist period (1945–1989) brought radical planning interventions (especially in the middle-sized and larger cities), which were motivated by the needs of the fast and large-scale industrialization and urbanization, and the persistent implementation of the Modernist zoning principles. The compactness of the urban structures was largely kept in that period due to the predominant role of public transport for mobility needs. Post-socialist urban development was strongly impacted by numerous economic and social factors, which resulted in urban restructuring, an increasing number of private cars, urban sprawl in the larger cities, and considerable shrinking in most of the smaller settlements.

The energy topic came to the focus of public attention in Bulgaria by mid 1980s due to shortcomings in energy supply that caused periodical setbacks in the urban energy systems. After 1990, with economic and social challenges emerging fast and the State largely withdrawing from the responsibility for energy production and supply, Bulgarian municipalities faced severe energy poverty. The burning issues in the urban context were mainly related to energy used for central heating and street lighting, electricity scarcity and unreliability, the growing bills, and the so-called electricity schedule.[3]

Capacity building for municipal energy planning was initiated as early as 1996 with international funding and methodological support. The continuous joint effort of numerous actors at the international, national, and local levels coordinated by the Centre for Energy Efficiency EnEffect,[4] resulted in the next decades in abundant experience and visible practical results. The Municipal Energy Efficiency Network *EcoEnergy* united by early 2003 overall 156 municipalities, accounting for about half the country population. Municipal energy efficiency (EE) offices were set up in the member municipalities and EE officers were nominated. Thus, a formal institutional network was gradually set up in the country, which acted as the organizational backbone of energy efficiency activities at the local level. Specialized software was designed and an analytical database of energy consumption in the Network members has been provided since 1997. The activities mobilized public support for local energy efficiency policy and stimulated important changes in public behavior. A model for integrating public policies with investment activities was successfully developed and tested. Active international contacts and good collaboration with the countries of Southeast Europe and with the European network Energy Cities were established. A Regional Network for Efficient Use of Energy and Water Resources (RENEUER) was set up in Southeast Europe (Nakova 2007).

Energy efficiency issues were among the targets of the practical implementation of the National Strategic Reference Framework of Bulgaria for the period 2007–2013[5] (MRDPW 2007) based on the national Regional Development Plan and the Operational Programme on Regional Development (2007–2013).[6] The major accent was there put on the development of various types of urban centers and the improvement of their

---

[3] Energy supply stopping daily in different parts of the city in alternative hours.

[4] A specialized non-profit NGO established in 1992 to support action towards developing energy efficiency policies in Bulgarian municipalities; http://www.eneffect.bg.

[5] Priority 4: Support for balanced territorial development.

[6] Part 3.8: Infrastructure development—energy infrastructure, including renewable energy sources and energy efficiency.

linkages with the adjacent rural regions. Transport infrastructure development was outlined as an important factor for the future spatial organization of the regions. The need for synchronization between regional and spatial plans and the cohesion aspects of development was a major focus in the recommended updating of the regional development plans at NUTS 2 level for the period 2009–2013 (MRDPW 2009). It was also recommended to explicitly outline the role of the plans in improving energy efficiency, in the implementation of renewable energy resources (RES), and in the reduction of greenhouse gas emissions. Yet, regretfully, there were no indicators on energy consumption and energy efficiency included within the recommended indicator set. A similar approach was applied to updating the mid-term planning documents—the District development strategies at NUTS 3 level for the period 2009–2013. It was expected that the access to quality services effectively provided by the transport and energy infrastructure would be analyzed in order to estimate changes since the beginning of the strategy implementation and to identify priority needs. Municipal development plans should, according the Guidelines, aim at identifying the common needs of neighboring municipalities, defining common priorities and coordinating joint action. The rationale behind that was the effort to concentrate existing capacity and resources, to improve EU funding accessibility, and to support territorial cohesion and local partnerships. The role of the municipal development plans was to be also estimated with regard to their role for increasing energy efficiency. Yet, there are no explicit energy accents in the recommendations on updating the SWOT analyses accompanying the plans. Neither was there such an accent traceable in the six strategic indicators chosen (Dimitrova 2012). The Annual Report on the implementation of the National Energy Efficiency Action Plan 2014–2020 (Republic of Bulgaria, Ministry of Energy 2016) provided a set of recommendations on future action based on the overview of observed tendencies. The recommendations were focused mainly on transport and industry systems. The need was claimed *"to incentivise final consumers, particularly households, to implement energy efficiency improvement measures,"* while the urban and regional planning levels were not addressed at all.

5.3 Strategic Spatial Planning at the Local Level in Bulgaria: Contributions to Energy Transition

Research Methodology

A comparative analysis was undertaken in late 2017 of the results of the authors' two recent research projects of different scope and focus: an academic project supported by the university research budget and focused on the evaluation tools developed for local strategic plans in Bulgaria[7]; and a multinational study funded under

[7] Research project entitled "Effectiveness of The Strategic Spatial Planning Instruments at Local Level: An Evaluation System," 2016–2017, partially funded by the university research budget, contract BN 182/2016.

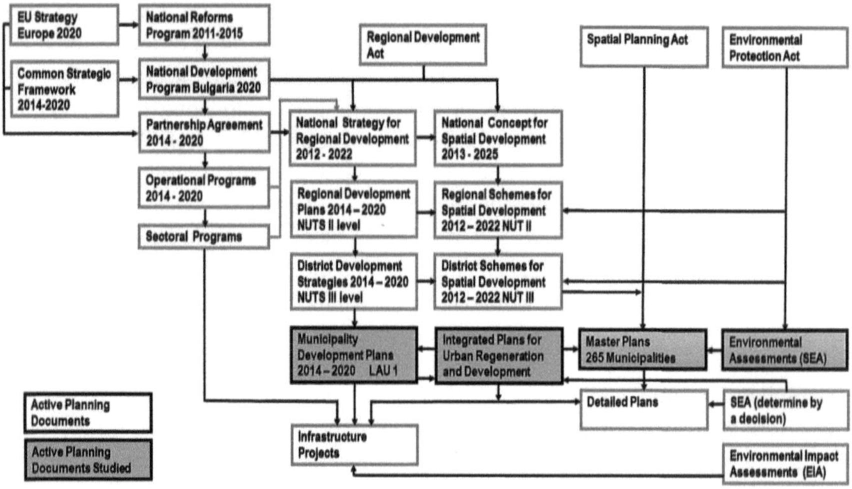

Fig. 5.1 The active national planning system of Bulgaria (the authors of the chapters)

Horizon 2020 Programme identifying the factors potentially enabling EU energy transition processes.[8] The study aimed to trace the existence of explicit and implicit considerations about energy efficiency as a sustainable development factor in the active strategic spatial planning documents at the local level in Bulgaria (Fig. 5.1). It searched for data complementarity in order to highlight two perspectives on the energy sensitivity of the changing Bulgarian planning system: from "inside out"—energy efficiency considerations reflected in the recently adopted local/municipal strategic spatial plans; and from "outside in"—societal appreciation of the role and capacity of the planning instruments of the local governance level in meeting energy-related challenges of the present day.

Research Steps Undertaken: The "Inside Out" Perspective

According to the active national legislation, the strategic local planning documents in Bulgaria comprise Municipal development plans (MDP), Integrated Plans for Urban Regeneration and Development (IPURD), and General Spatial Plans of the municipalities (GSPM) with Strategic Environment Assessments (SEA) (Fig. 5.1).

The study of the strategic planning documents at the local level focused on the planning documents concerning 18 Bulgarian municipal centers included in the Statistics on European cities[9] *with data available from the Copernicus Land*

[8] ECHOES project, (Energy CHOices supporting the Energy Union and the SET plan), 2016–2019, funded by HORIZON 2020 Programme, Contract No: 727470, Call H2020-LCE-2016-2017.

[9] Previously known as Urban Audit cities.

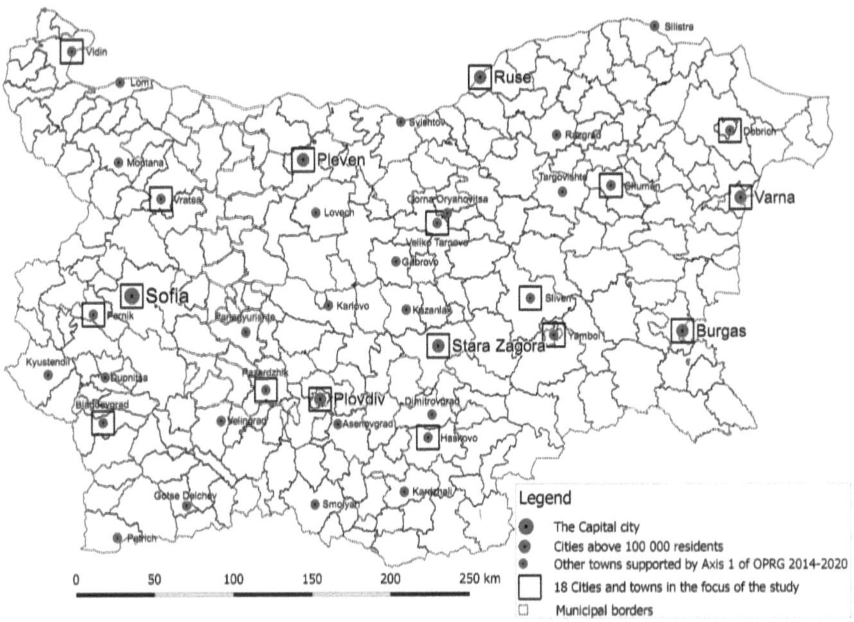

Fig. 5.2 Geographical location of cities with Integrated Plans for Urban Regeneration and Development (IPURD) in Bulgaria included in the study (the authors of the chapters)

Monitoring Service[10] *(including the Urban Atlas data sets for the Functional Urban Areas of these cities). The selection took into consideration the availability of statistical and geo-spatial data sets, providing baseline data and good ground for comparison between the midterm and final evaluation of the impacts from the implementation of the plans. The selected cities also belong to a larger group of 39 cities supported through Axis 1of the "Regions in Growth" Operational Program 2014–2020 for the development of integrated plans for urban regeneration and development* (MRDPW 2011) (Fig. 5.2).

The evaluation criteria applied concerned the availability of published documents; the presence of formulated energy efficiency-related goals and impact indicators; the level of correspondence between strategic goals and impact indicators; and the level of goal aggregation. Five evaluation grades were applied[11] (Fig. 5.3).

[10] Copernicus is a European system for monitoring the Earth. Data is collected by different sources, including Earth observation satellites and in situ sensors. The data is processed and provides reliable and up-to-date information in six thematic areas: land, marine, atmosphere, climate change, emergency management, and security. The land theme is divided into four main components: global, pan-European, local, references: https://land.copernicus.eu/.

[11] The five grades comprise: 0—missing goals and impact indicators, or unavailable published documents; 1—missing goals and impact indicators with appropriate focus on energy efficiency; 2—the goals related to energy efficiency are poorly addressed or they are formally mentioned without relevant impact indicators or indicators for resources or results; 3—formulated goals in the field of energy efficiency, mismatch with the impact indicators; 4—formulated energy efficiency

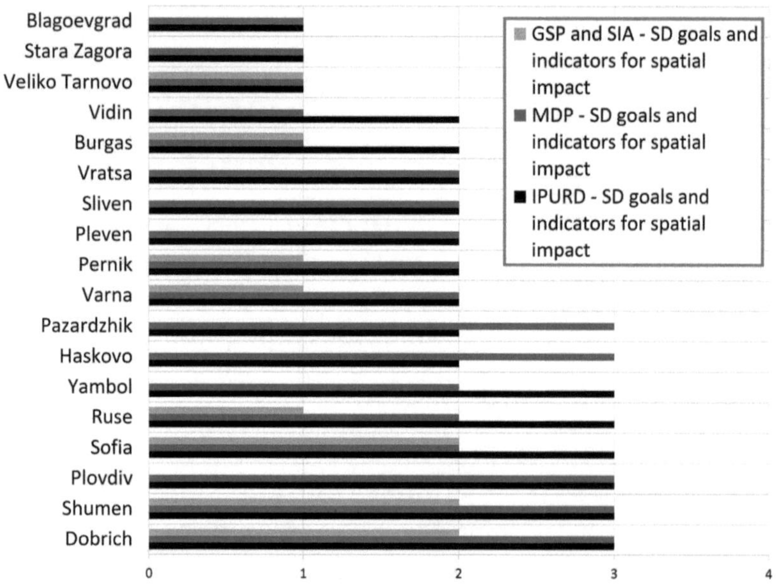

Fig. 5.3 Expert evaluation of the active strategic local plans in the 18 municipalities (authors)

The analysis identified 10 out of 18 municipalities with lacking active or draft GSPM—neither Internet-accessible nor published in the webpage of the municipality; only 3 municipalities (Sofia, Dobrich, and Shumen) and planning documents that were ranked as average/satisfactory (rank 2). Among the 18 IPURD analyzed, only six (Sofia, Plovdiv, Ruse, Dobrich, Shumen, and Yambol) ranked higher than average (Rank 3); three IPURDs (Stara Zagora, Blagoevgrad, and Veliko Tarnovo) were ranked below the average score (Rank 1). The MDP analysis identified five municipalities (Plovdiv, Dobrich, Shumen, Pazardvik, Haskovo) with higher than the average score (Rank 3). The evaluation of IPURD goals and indicators outlines the relatively low score for most of the plans where goals and/or impact indicators for energy efficiency were either missing or unconvincingly defined.

The content analysis of the IPURD followed three research steps: firstly, a sustainable development conceptual framework was chosen to outline major imperatives and tensions related to the concept and the resulting goal lines of action envisaged by the planning documents. Secondly, the expected focus of relevant policy targets and indicators addressing the particular SD aspects was determined through expert evaluation. Thirdly, the content of the selected plans for urban regeneration and development was analyzed in order to identify explicitly or implicitly present energy-related implications in the documents. The topic of energy and energy efficiency was searched for through content analysis of the declared goals, actions, and sets of considerations (Fig. 5.4).

---

related goals with corresponding impact indicators.

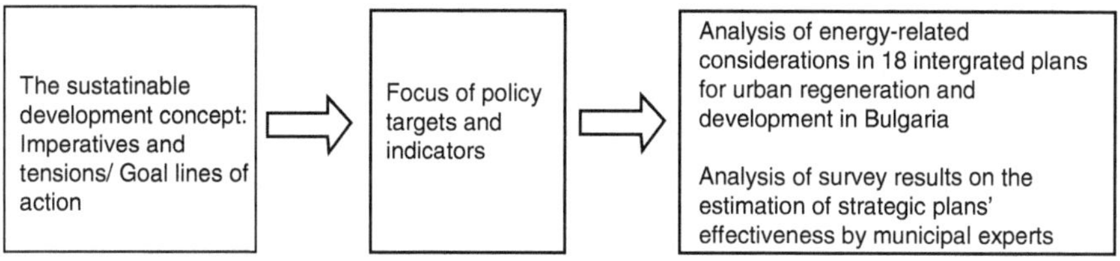

Fig. 5.4 Methodological concept of the content analysis (authors)

The conceptual frame developed for the purpose of the study was based on Campbell's identification of SD topics and contradictions in the urban planning field (1996) and Spangenberg's Prism of Sustainability (Valentin and Spangenberg 2000), setting SD imperatives at the four points of a tetrahedron and outlining the major tensions among them along the four edges linking the points. The sets of action goals in the planning documents were related to the need for meeting the demand of (sometimes mutually contradicting) imperatives, namely: (a) environmental (decrease of pressure on the environment and the resources); (b) economic (increasing competitiveness and entrepreneurship); (c) social (cohesion and equality); and (d) institutional (good multilevel governance) ones. The energy efficiency topic as part of the broader area of resource and energy efficiency targets explicitly appears in the model as a challenge in addressing the tension between the environmental and economic imperatives.

The content analysis of the IPURDs outlined the presence of both explicitly and implicitly mentioned energy efficiency considerations (Annexes 1 and 2). The plans rather lack a holistic view on energy efficiency, behavior, performance, transformation, etc. Energy efficiency considerations are mostly present through general statements on intended action and targeted outcomes. The specified resource remain broadly not quantified—a common challenge for all the dimensions due to the lack of appropriate information or inadequate measurements. The major focus in terms of measures and projects is on multifamily residential buildings and on public ones, which takes into account the reference framework at the national level. Nine documents out of the 18 analyzed ones refer to the environmental imperative and the corresponding action goal for decreasing the pressure on environment and resources. The reduction of CO_2 emissions is explicitly mentioned among the indicators, yet without specific baseline figures or targets with clear territorial reference. Reduced energy consumption is also mentioned—once only generally and once with the only quantitative indicator introduced ("number of households with improved energy consumption"). One plan includes an indicator of "increased share of electric urban transport." The social imperative is mainly addressed through action aimed at improving the quality of life through "diversification and accessibility of social services" and by providing "efficient public services." The renewal and expansion of public service facilities and the improvement of the municipal social infrastructure, mentioned in several plans might be expected to have implicit considerations about improving energy efficiency. The action goals addressing the economic imperative are mainly focused on competitiveness and knowledge-based economy. Smart, green, and inclusive growth is expected to contribute to increasing the attractiveness and the competitiveness in the regional and national context. The increased energy efficiency of buildings is explicitly recognized as an important driving factor for economic development. The institutional imperative is logically related to multilevel governance, which is seeking to implement integrated urban management for better land use, mobility, public services, and energy-efficient infrastructure. In most of the declared IPURD goals, the tension between institutional and social imperatives is addressed by using action verbs that frame and channel energy transformation—raising awareness rand motivation, promoting energy efficiency and environmental responsibility, support for implementing energy efficiency measures. The tension between institutional and environmental imperatives is addressed in the plans by action aimed at the preservation of natural and cultural assets; energy efficiency is not explicitly mentioned there among the considerations. The action envisaged at different levels of intervention and territorial scope mainly focuses on revitalization, conservation, sustainable development of resources, regeneration, restoration, and integrated urban development. The tension between **environmental and economic imperative** is the most frequently addressed one; it is mentioned in 11 out of the 18 plans. Planned action focuses explicitly on resource and energy efficiency—it includes improvement of the technical urban infrastructure, public and residential buildings. The tensions between **social and economic imperative** and between the **institutional and economic imperative** are the least mentioned ones and only generally addressed by the documents. The issues with spatial implications there comprise improving the quality of the public and working physical environment and the accessibility of educational and cultural sites. Overcoming imbalances between different areas in the city is claimed a priority in one document.

The survey undertaken within the university research project[12] aimed to identify experts' estimation of the availability of up-to-date strategic planning instruments at the municipal level and the existing capacity for strategic spatial planning. The three types of planning documents were focused upon—master plans, municipal development plans, and IPURD. The questions addressed the extent of coordination between the plans; the accessibility of relevant information and technologies; the level of local expertize especially with regard to monitoring and evaluation procedures. The survey covered all types of municipalities in the country and a variety of positions and expert fields. The analysis of results outlined important peculiarities of the current situation at the local level: The strategic planning documents in over 80% of the municipalities have been either adopted within the last 10 years or are currently under development. Regulatory requirements, observations, and evaluations on ongoing development processes and investment pressure are the main reasons claimed for undertaking changes of the planning instruments. In less than 1/3 of the responding municipalities, some kind of publicly accessible tracing of the implementation process has been introduced. The advance in IPURD implementation is mainly traced by publishing data about the implementation of the investment projects. *The coordination between*

[12] Questionnaires sent to the 39 municipalities eligible for IPURD funding, response received from overall 15 municipalities.

the different plans in time and space *were* estimated as good by only one third of the respondents and almost half (43%) consider them "instable." Less than half of the municipalities provide publicly accessible data with spatial dimensions information is internally provided within the municipalities in various ways, yet only one municipality implements GIS-based information exchange. All the respondents maintain a register of the municipal investment projects included in the IPURD yet only in 21% it is publicly accessible. Similarly to other plans, GIS-based instruments are implemented in only 14% of the responding municipalities. About 1/3 of the municipalities generate data on the separate project implementation but without referring to indicators; about ¼ of the municipal administrations do not use the generated information in any way. The responding municipalities have either established separate managerial structures (with mainly experts in economics and EU funding projects) or some working coordination mechanism with urban planning experts involved. GIS-based working environment is still missing in 50% of the respondent municipalities. Respondents largely claimed the need for additional expertise in the field of local monitoring and evaluation in the strategic planning field, recommending the establishment of a specialized regional or inter-municipal center. The need for improved expertise and additional education at the municipal level was largely claimed.

Research Steps Undertaken: The "Outside-In" Perspective

Being a multi-disciplinary research project at EU level, ECHOES project is aimed at providing policy-makers with comprehensive information, data, and policy-ready recommendations about the successful implementation of the Energy Union and SET plan. Individual and collective energy choices and social acceptance of energy transitions are analyzed in a multi-disciplinary process including key stakeholders as co-constructors of the knowledge. To account for the rich contexts in which individuals and collectives administer their energy choices, the ECHOES keeps and explicit research focus on people's experience, modes of behavior, feelings and attitude to public policies, policy instruments, etc. That has determined the research approach and the instruments applied—focus groups and interviews with various actors.

The three preliminary focus groups conducted in Bulgaria in the period April–November 2017 searched for the points of view of three main decision-making levels addressed by ECHOES project—municipal,[13] national,[14] and collective[15] ones.

[13] Focus group in Gabrovo, April 2017; six participants from the expert level (municipalities).

[14] Focus group in Sofia, July 2017; six participants from institutions working at the national level: Ministry of Regional Development and Public Works; Bulgarian Academy of Sciences; three NGOs.

[15] Focus group in Sofia, November 2017; six participants from the level of collective decision-making covering diverse situations in terms of: (a) role in the building management process; (b) period of experience as heads of the management boards; (c) professional background, and (d) age, construction, function, and stage of maintenance of the building.

All the participants were invited to answer six questions: (Q1) In what role have you participated in the energy transition process in your institution/organization? (Q2) What were the motivating factors for undertaking/not undertaking action in the three fields focused upon—buildings, e-mobility, and smart technologies? (Q3) What were for you (a) the main achievements and (b) barriers in the process? (Q4) Which was the most satisfying result for you? Why? (Q5) What was your biggest disappointment? Why? (Q6) What would you recommend in order to increase the effectiveness of the action undertaken? The feedback received through the three focus groups provided insights on participants' perceptions and estimations on the energy transformation process in Bulgaria from different points of view.

The important role of municipalities in the process towards greater energy efficiency in Bulgaria was acknowledged in all the three working groups. Energy efficiency in residential buildings was outlined as a particular priority due to the estimated poor conditions in the aging prefabricated multifamily residential buildings and the urgency of upgrading them. Despite expressed criticism, the legal framework, the national policy framework, and the national strategy in action alongside the availability of EU funding were considered important factors for undertaking action. Personal commitment in the process was declared by all the respondents. Satisfaction was expressed with the visible positive changes occurring in the urban environment as a result of the already accomplished refurbishment projects and a serious concern about the long-term effects of the action. There was a general understanding about the need for integrated solutions; NGOs claimed the need for a long-term vision, transparency, and continuity of implementation. Yet, there was no reference to the local strategic plans as possible tools for addressing energy efficiency at the local level.

Ten in-depth semi-structured interviews were undertaken in August and September 2017—three with representatives of formal social units[16] and seven with representatives of the collective decision-making level.[17] The subjects most frequently discussed were clustered around several keywords: buildings, projects, investments; electricity, heating, rehabilitation/insulation, energy consumption, energy efficiency; policy and action levels. Several interrelated topics were encountered with high frequency in all the responses, among them the decentralization of energy production and the role of municipalities in the energy transition process. Not a single mention of spatial planning as an active or potential factor in the process was registered, yet the local level of governance was mentioned in relation to energy costs; market mechanisms versus political regulations in the framework of sustainability, raising increasing energy efficiency awareness, and an increasing capacity to act.

[16] Representatives of Ministry of Energy, Sustainable Energy Development Agency and an Electricity System operator.

[17] Representatives from Sofia Municipality (decision-making level), three private companies and three NGOs.

Discussion of Results

The undertaken content analysis of the information generated through university research project provided an insight on the degree of awareness, understanding, and capability of dealing with energy issues in the planning field at the local governance level; the survey among municipal planning experts outlined expert perceptions and estimations on what is being done, what is needed, and what capacity is still missing for strategic planning at the local level. The study confirmed that energy efficiency is already explicitly addressed by the strategic planning documents at the local level in Bulgaria, the major reason being the support for the topic by published official EU documents and by the availability of EU funding. Energy efficiency was strongly focused upon by the Integrated plans for urban regeneration and development (IPURD), funded through the Operational programmes where funding claims for energy efficiency measures had an explicit priority. Projects meeting the needs in residential units with large built area and considerable number of inhabitants, and in administrative buildings were eligible for funding through the axis addressing urban development in the *Regions in Growth* Operational programme. That was the reason why targets and indicators related to improving the energy efficiency of residential and administrative buildings were included in all the integrated plans already developed. Due to the requirement for explicitly linking the integrated plans for urban regeneration and development with the municipal development plans, energy efficiency appears in these local documents as well. The analysis of the indicators defined for the needs of evaluating the results and impacts of the IPURDs indicates that by the beginning of the second funding period for Bulgaria (2014–2020) most of the local authorities measure the results and impacts by the number of buildings with implemented EE measures with broadly differing understanding about the type and number of measures. Even if some indicators have been defined for measuring the amount of energy saved, the lack of baseline indicators and particularly defined targets is indicative for the need of much broadly promoting the topic of EE and of providing technical expert support; that would be expected to help in the process of conceptualizing the link between the two topics—resource and energy efficiency, and sustainable development. Having in mind the strong influence of national policies on local level decision-making in Bulgaria, visibly exerted also through the access to funding, the integration of various policies (including EU policies as well), should be continually enriched in future at the national level as well. Considerable diversity among municipalities in Bulgaria should also be explicitly mentioned with regard to their preparedness to act for energy transition. Some of them are still lagging behind, others have persistently implemented their EE policy for many years, successfully including resource and energy efficiency in their IPURD—a convincing example could be discovered in the municipality of Gabrovo, one of the co-founders of the *EcoEnergy* Municipal Network for Energy efficiency in Bulgaria (Dimitrova et al. 2014).

Two major conclusions were made: (1) Insufficient integration of different strategic local plans, which hampers their mutual integration; (2) The lack of clear regulatory basis on evaluation and monitoring instruments results in lacking or ineffective practice of evaluating the implementation of the local plans (excluding environmental impact assessment).

There is important work still to be done with regard to the recommended development of a clear system for monitoring and evaluation of the spatial plans implementation. Based on the survey results, it could be expected that Bulgarian municipalities would be motivated to enter in such a process. The development of relevant procedures and instruments would however require close preliminary work with all the stakeholders for solving problems of land ownership, maintenance and use, and establishing clear rules for coordination between different strategic plans and with the operational plans. Additional education and training requested at the municipal level, establishment of regional centers to provide methodological support in introducing GIS-based spatial analyses.

A growing awareness about the complexity of the development process and the leading role of municipalities was repeatedly demonstrated during all the focus groups and interviews undertaken within ECHOES project. Many of the interviewed and focus group participants believe that joint decisions taken at an expert and political level provide the relevant basis for action; call for a long-term vision, transparency, and continuity of its implementation no matter of political party into power and future political changes. Urgent needs identified are concerning interrelated changes in national legislation and regulations in order to link energy-related issues to issues of spatial planning and design, real estate property, and public procurement. With respect to socio-cultural identity and regarding economic and geographical contexts, much more attention should be channelled towards country, regional and place specific approaches to energy transition, urban planning, and management of the built environment. Social aspects of the process were high on the agenda. Capacity building was another key and urgent consideration of most of the participants. The process requires a lot of information, which is difficult to collect and continuously update; huge financial and human resources; considerable expert capacity to build; and broader opening of the administration to external experts and NGOs.

The active national legal frameworks, in general, and the acts directly related to energy and energy efficiency, in particular, were often mentioned as enablers as well as barriers in the process. Other acts with direct impact on local development are the Concessions Act, the Public Procurement Act, the Condominium Act, etc. The more complicated process of applying for funds, changing rules and delayed review procedures of the submitted projects, the more barriers are emerging—administrative, technical, and financial. The uncertainty among households and organizations (include lack of knowledge about and confidence in new technology and business models) results in replicating established models and solutions.

Several factors acting as motivators and/or barriers in the relationship between different decision-making levels should be outlined: joining in EU projects and networks provide a general breakthrough in municipalities for undertaking action in the energy efficiency field. National regulations play important role in the process as a framework to follow and further upgrade. Relevant professional expertise at the local level positively influences both personal motivation for action and policy implementation as a whole; yet incentives and methodological support from the national level are expected on behalf of the municipalities.

The self-reflexion on efforts and results could be more broadly applied in realizing the need for more intelligent approaches and integrated solutions. The already accumulated practical experience contributes for greater self-confidence and additional motivation for going on and doing things together. Furthermore, visible changes and results in the overall image of the urban environment because of rehabilitation are equally important to reducing particular energy expenditures, but also to encouragement for action, results multiplication at various locations, achieving synergies through concentrated actions and projects. The lack of political will and understanding at local level could be considered a barrier most difficult to overcome. In addition, the lack of coordination between institutions brings fragmented results and lack of continuity in the process. The complementarity of bottom-up initiatives and top-down measures could be considered most effective in achieving sustainable results, as most of the energy infrastructure and the projects require decisions at regional, national, and even higher level in order to provide a common European energy transformation process. An implicit message to the planning system is the request that the local level of public governance uses a clear language on sustainability priorities when interpreting and implementing EU policy/programmes.

5.4 Conclusion

The identified general messages to the governance and expert levels comprise: (1) need for further practice-oriented research on the definition of working instruments for monitoring and evaluation of the plans effectiveness; (2) need for elaboration of societal debate on clear definition of the goals and values of urban development and planning; (3) the need for definition of the scope and the structure of the required information but also the self-sustained mechanisms for its generation, update, and analysis; (4) needs for outline and guidelines on the integration of the evaluation into the decision-making process; (5) needed interdisciplinary and transdisciplinary approaches which are yet underdeveloped; 6) need for development of the potentials of the expert capacity under the leading role of the formal and non-formal education.

After the political changes in 1990s, the local level of governance in Bulgaria was attributed a lot of responsibilities with often lacking institutional capacity and technical expertise for meeting them and with practically lacking or minor decentralization of financial resources. During the recent three decades, many municipal authorities in the country were actively taking chances—through EU funding programmes, national institutional and expert support, and international networks; they proved to be inventive and working hard in the field of energy efficiency, looking for funding but also for partners among a broad variety of stakeholders. Alongside economic considerations, social concerns are continuously high on the political agenda of Bulgarian municipalities because of the complex societal context—aging and energy-inefficient residential and public service buildings in urgent need for refurbishment, comparatively high levels of unemployment and low income for major part of the population and lagging economic development in mostly shrinking settlements. The effort proved fruitful in many cases; it already resulted in visible positive changes in the urban environment of the active municipalities; it is also successful there in building local expertise, changing mindsets, and drawing attention to the need for strategic thinking, a long-term vision, and integrated approaches among all actors and institutions involved in the local process of energy transformation. Nowadays, the leading role and the capacity of the public sector at the local level of governance in the Bulgaria process is broadly acknowledged at all levels—from the EU, to the national and local ones.

The energy topic has already entered the field of strategic spatial planning in Bulgaria and has been acknowledged as an important one in addressing the sustainable development imperatives at the local level of governance. The currently missing clear and holistic general framework for monitoring and evaluation and the acknowledged need for building it should be considered a good chance for undertaking action, also explicitly integrating the energy dimensions of the urban process. There is, in parallel, an urgent need for interdisciplinary research and joint development of intensive interdisciplinary educational and training programmes in support of municipalities.

Despite being present in EU research field and in governance practice elsewhere, the role of spatial planning in the process of energy transformation is still strongly underestimated and regarded with a rather narrow focus on buildings refurbishment at all governance levels in the country. Although not explicitly mentioned as an enabling factor or a barrier during the focus group discussions, it could be however linked to claims made for the necessity of strategic thinking and integrated approaches to be applied at the local level of governance. Research has also identified motivation and positive experience in some Bulgarian municipalities in integrating energy considerations in strategic spatial planning instruments. It is a tendency that surely needs methodological and institutional support from higher levels and more broadly introducing the topic into public debate and educational programmes. Building awareness about the importance and benefits of linking energy to municipal strategic planning could be expected to further mobilize local effort and would be expected to result in motivation for developing and locally applying relevant tools for evaluating the effectiveness of spatial planning in terms of energy transition. That would provide the public sector at the local level with a powerful instrument for coordinating local efforts to economically, socially, and environmentally sustainable territorial development.

Annex 1: Explicit and Implicit Responses by *IPURD* to Sustainable Development Imperatives

SD concept: imperatives/ action goals	Focus of related policy targets and indicators measuring progress	Explicit and implicit energy-related considerations in the selection of 18 plans for urban regeneration and development in Bulgaria (number of documents referring to particular topic)
Environmental imperative/decreasing pressure on environment and resources	Material and energy flows, emissions, and waste from or towards different territorial location and their impact on the environment	Reduction of greenhouse gas emissions (1); Reduction of carbon emissions (4); Reduced energy consumption in renovated administrative and residential buildings (1); Reduced consumption of final and primary energy in dwellings and administrative buildings (1); Number of households with improved energy consumption (1); Increased share of electric urban transport (1)
Social imperative/social cohesion and equity	Level of satisfaction with quality of life, the built environment, and public services	Turning into a commercial, spiritual, and educational center (1); Development of sites of high public significance (1); Development and delivery of efficient public services (1); Renewal and expansion of public service facilities in the intervention zones (1). Enhanced quality of social services—diversification and accessibility (1); Improvement of the municipal educational, health and social, cultural and sports infrastructure (2); Integrated indoor and outdoor living environment providing high quality of life (1); Improving quality of life (1); Sustainable integrated development based on raising the quality of the living environment (1)

(continued)

SD concept: imperatives/ action goals	Focus of related policy targets and indicators measuring progress	Explicit and implicit energy-related considerations in the selection of 18 plans for urban regeneration and development in Bulgaria (number of documents referring to particular topic)
Economic imperative/increasing competitiveness and entrepreneurship	Level of innovation activities, research and development on site, and the local, city, and regional value-added impacts	Turning into a leader with attractive investment environment with conditions for growth (1); To transform the city into a research and cultural center (1); A sustainable city with competitive productions (1); Achieving sustainable knowledge-based economic growth (1); Building a competitive economy and information society (1); Scientific progress and economic development (1); Stimulating the economic development by improving the energy efficiency of buildings (1); Improving the city's attractiveness and competitiveness in the regional and national context by smart, green and inclusive growth (1)
Institutional imperative/good multilevel governance	Level of citizens' satisfaction with availability and access to infrastructure providing public services; improved quality of life; and built environment	Turning the city into a Black Sea leader with high quality of life (1); Efficient governance (1); Integrated urban management of areas with negative development trends and unrealized development potential (1); Effective management and tourism sustainable development (1); **Improvement of the main city infrastructure through investment in energy efficiency, including, integrated transport infrastructure** (2); Sustainable and energy-efficient development of the city and its functional systems (1); **Provision of environmentally sustainable and efficient public services (1)**; Enhanced quality of social services—construction and reconstruction of facilities (1); Development of efficient public services (1)

Annex 2: Explicit and Implicit Responses by *IPURD* to Tensions Between Sustainable Development Imperatives

SD concept: tensions between imperatives/action goals	Focus of related policy targets and indicators measuring progress	Explicit and implicit energy-related considerations in the selection of 18 plans for urban regeneration and development in Bulgaria (number of documents referring to particular topic)
Institutional—social tension imperative/ increased participation and improved transparency	Level of communities' representation, integration, and cooperation; anticorruption measures; accountability and empowerment, level of civic control	Supporting the implementation measures for energy efficiency and environmental responsibility (1); Support and stimulation for energy-efficient consumption and renewable energy use in the built urban environment (1); Stimulating energy efficiency and renewable energy sources (1); Promoting energy efficiency in buildings and installations and the deployment of renewable energy sources (1) Stimulating social integration (1); Creating conditions for overcoming the imbalance between different areas in the city (1)
Institutional vs. environmental tension imperative/ preservation of natural and cultural assets	Effectiveness of administrative procedures and/or imperfections and gaps related to the spatial organization of cities and competing land use	Revitalization; improvement of the local conditions (1); Ensuring high quality of life; conservation and development of the natural and built environment (1); Conservation and efficient use of natural resources (1); Sustainable ecological development and local resource conservation (1); Green, environmentally friendly urban regeneration (1); Restoration and development of the living environment and increasing energy efficiency (1); Revival of the city by restoration and development of the urban environment (1). Sustainable integrated urban development based on local potential and improved economic, social, and territorial impact of the developed zones (1)
Environmental vs. social tension/ resilience to burden and change	The ecological and spatial footprint of communities' activities, their adaptability, and financial indebtedness	Improved living environment, clearly recognizable cultural and spiritual values, integrated into the life of local people (1)

(continued)

SD concept: tensions between imperatives/action goals	Focus of related policy targets and indicators measuring progress	Explicit and implicit energy-related considerations in the selection of 18 plans for urban regeneration and development in Bulgaria (number of documents referring to particular topic)
Environmental vs. economic tension/ resource and energy efficiency	The energy and resource capacity of morphologically specific and spatially defined units of production and consumption	Efficient resource use and provision of quality public services (1); Improving the energy efficiency of the building stock, meeting modern living standards with reduced energy consumption (4); Introducing energy saving technologies and alternative energy sources in public buildings (1); Reducing the energy dependence of public buildings (1); Stimulating the economic development by improving the energy efficiency of public buildings (1); Improving energy efficiency in public and residential buildings (1); Energy efficiency of buildings and installations and the deployment of renewable energy sources (1); Improvement of the technical infrastructure and the quality of the living environment (1)

References

Bertaud A (2004) Thespatial structures of Central and Eastern European cities: more European than socialist? In: International symposium on post-communist cities, University of Illinois at Urbana-Champaign, 17–19 June 2004. http://alainbertaud.com/AB_Files/AB_Central%20 European%20Spatial%20Structure_Figures_2.pdf. Accessed 23.04.2018

Biresselioglu ME, Burov A, Carrus G, Demir MH, Dimitrova E, Fritsche I, Kaplan MD, Klöckner CA, Koksvik G, Masson T, Mutafchiiska I, Panno A, Røyrvik J, Schwarzinger S, Skjølsvold TS, Tasheva M, Tiberio L, Velte D, Vesely S (2017) Social science perspectives on electric mobility, smart energy technologies, and energy use in buildings—a comprehensive literature review, ECHOES Report NECHOES 3.1.-D3.1 Literature Review

Campbell S (1996) Green cities, growing cities, just cities? Urban planning and the contradictions of sustainable development. J Am Plan Assoc 62(3):296–312. American Planning Association, Chicago, IL

Davoudi S (2010) Why territory matters for competitiveness and cohesion? In: Presentation, Network Conference 2010 on Territorial Cohesion in Central, Eastern and South Eastern Europe—Challenges ahead for Strategic Planning and Urban-Regional Governance, Belgrade, Serbia, 22–24 Sept 2010. http://www.spa-ce.net/pdf/2010/spa-ce.net%202010%20-%20 Davoudi.pdf. Accessed 23.04.2018

Dimitrova E (2007) Testing PETUS: expectations and outcomes of the theory—practice dialogue on urban sustainability. Indoor Built Environ 16(3):216–225. International Society of the Built Environment, SAGE Publications

Dimitrova E (2012) Socio-spatial dimensions in sustainable resource and energy management in Bulgaria. In: Proceedings, 26th AESOP annual conference, METU, 11–15 July 2–12, Ankara

Dimitrova E, Burov A (2010) Current challenges and innovative approaches in the development of peripheral regions in Bulgaria (the case of Ivaylovgrad municipality). Spatium Int Rev 23:1–8. Belgrade, Serbia

Dimitrova E, Burov A, Nakova K (2014) Bulgaria. In: Jones P et al (eds) COST Action TU1104—smart energy regions. The Welsh School of Architecture, Cardiff University, pp 33–43

EC (2007) Leipzig Charter on Sustainable European Cities. Adopted by the Informal Ministerial Meeting on Urban Development and Territorial Cohesion. www.eufunds.bg/document/355. Accessed 23.04.2018

EC (2010) Toledo Declaration. Adopted by the Informal Ministerial Meeting on Urban Development in Toledo, 22 June 2010. http://urban-intergroup.eu/wp-content/files_mf/es2010itoledodeclaration.pdf. Accessed 23.04.2018

EC (2011) Territorial Agenda of the European Union 2020. Towards an Inclusive, Smart and Sustainable Europe of Diverse Regions. Agreed by the Informal Ministerial Meeting of Ministers responsible for Spatial Planning and Territorial Development on 19th May 2011 in Gödöllő, Hungary. http://ec.europa.eu/regional_policy/en/information/publications/communications/2011/territorial-agenda-of-the-european-union-2020. Accessed 23.04.2018

EC (2012) Directive 2012 on Energy Efficiency, amending Directives 2009/125/EC and 2010/30/EU and repealing Directives 2004/8/EC and 2006/32/EC. European Commission, Brussels

EC (2015) Towards an Integrated Strategic Energy Technology (SET) Plan: accelerating the European Energy System Transformation. Communication from the Commission C (2015) 6317 final, 15.9.2015, Brussels. https://ec.europa.eu/energy/sites/ener/files/documents/1_EN_ACT_part1_v8_0.pdf. Accessed 23.04.2018

EC (2016a) Good Practice in Energy Efficiency: for a sustainable, safer and more competitive Europe. REMOURBAN project (Regeneration MOdel for accelerating the smart URBAN transformation), Smart City Network. http://www.remourban.eu/Technical-Insights/Insights/Good-Practice-In-Energy-Efficiency.kl. Accessed 23.04. 2018

EC (2016b) Urban Agenda for the EU ('Pact of Amsterdam'). Agreed by the Informal Meeting of EU Ministers Responsible for Urban Matters on 30 May 2016 in Amsterdam, The Netherlands. http://ec.europa.eu/regional_policy/sources/policy/themes/urban-development/agenda/pact-of-amsterdam.pdf. Accessed 23.04.2018

EC/COM (2001) 'A sustainable Europe for a better world: a European Union strategy for Sustainable Development' (Commission proposal to the Gothenburg European Council) [COM(2001) 264 final]. https://eur-lex.europa.eu/legal-content/EN/TXT/PDF/?uri=CELEX:52001DC0264&from=EN. Accessed 23.04.2018

EC/COM (2005) Review of the sustainable development strategy—a platform for action [COM(2005) 658 final]. https://eur-lex.europa.eu/legal-content/EN/TXT/HTML/?uri=LEGISSUM:l28117&from=EN. Accessed 23.04.2018

EC/COM (2011) Energy Roadmap 2050. Communication from the Commission to the European Parliament, the Council, the European Economic and Social Committee and the Committee of the Regions. http://ec.europa.eu/energy/energy2020/roadmap/doc/com_2011_8852_en.pdf. Accessed 23.04.2018

EEA (2001) Designing effective assessments: the role of participation, Science and Governanceand Focus. Environmental Issue Repost No. 26

EEA (2006) Urban sprawl in Europe—the ignored challenge. Report No 10/2006. https://www.eea.europa.eu/publications/eea_report_2006_10/eea_report_10_2006.pdf/view. Accessed 23.04.2018

EEA (2013) Achieving energy efficiency through behaviour change: what does it take? EEA Technical report No 5/2013

EEA-FOEN (European Environment Agency and Federal Office for the Environment) (2016) Urban sprawl in Europe: Joint report

Efficiency First Foundation (2016) Driving competitiveness, energy security and decarbonisation through increased energy productivity

ESDP (1999) European spatial development perspective. towards balanced and sustainable development of the territory of the European Union. Agreed at the Informal Council of Ministers responsible for Spatial Planning in Potsdam, May 1999

Girard LF, Cherreta M, de Toro P (2007) Integrated panning and integrated evaluation—theoretical references and methodological approaches. In: Miller D, Patassini D (eds) Beyond benefit cost analysis. Accounting for non-market values in planning evaluation, urban planning and environment, Ashgate

Healey P (1997) Collaborative planning: shaping places in fragmented societies. Macmillan, London

ICLEI (1995) Charter of European Cities and Towns "Towards Sustainability". (The Aalborg Charter). European Conference on Sustainable Cities & Towns. http://www.sustainablecities.eu/fileadmin/repository/Aalborg_Charter/Aalborg_Charter_English.pdf. Accessed 23.04.2018

Institute for Energy and Transport (2012) Analysis of energy saving potentials in energy generation: final results. Publications Office of the European Union, Luxembourg

Khakee A (2003) The emerging gap between evaluation research and practice. Evaluation 9(3):340–352. SAGE Publications

Miller D, Patassini D (eds) (2007) Beyond benefit cost analysis. Accounting for non-market values in planning evaluation, urban planning and environment, Ashgate

MRDPW (Ministry of Regional Development and Public Works) (2007) Operational programme 'Regional Development' 2007-2013. DG 'Programming of Regional Development', Sofia, http://www.bgregio.eu/en/. Accessed 23.04.2018

MRDPW (2009) Methodological guidelines for the updating of active regional and local development strategies and plans, General Directorate for Regional policy and governance systems, Sofia (in Bulgarian). http://www.mrrb.government.bg/static/media/ups/articles/attachments/85e5bca10edf4bce39cc99f7f87d6776.pdf. Accessed 23.04.2018

MRDPW (Ministry of Regional Development and Public Works) (2011) Opening procedures on contracts under the support for integrated plans for urban regeneration and development scheme. News, 22 June. https://www.eufunds.bg/archive/documents/1310454633.pdf. Accessed 23.04.2018

Naess P (2006) Urban structure matters. Residential location, car dependence and travel behaviour. The RTPI Series. Routledge, London and New York

Nakova K (2007) Energy efficiency networks in Eastern Europe and capacity building for urban sustainability. The experience of two municipal networks. Indoor Built Environ 16(3):248

Olazabal M, Pascual U (2015) Urban low-carbon transitions: cognitive barriers and opportunities. J Clean Prod 109:336–346

Partnership on Energy Transition (2017) Orientation Paper—Energy Transition. https://ec.europa.eu/futurium/en/system/files/ged/3.orientation_paper_energy_transition.pdf. Accessed 28 Mar 2018

POINT Project (2007) (ID: 217207, Funded under: FP7-SSH), How effective are policy indicators?

Republic of Bulgaria, Ministry of Energy (2016) Annual Report on the implementation of the National Energy Efficiency Action Plan 2014–2020, Sofia

REViSITE Coordination Action (2012) www.revisite.eu. Accessed 31.3. 2018

Sen S, Ganguly S (2016) Opportunities, barriers and issues with renewable energy development—a discussion. Renew Sust Energ Rev 69:1170–1181. Accessed 6.10. 2016

UN (United Nations) (1992) Earth summit. Agenda 21. https://sustainabledevelopment.un.org/content/documents/Agenda21.pdf. Last accessed 23.04.2018. Accessed 23.04.2018

UN (2015) The Sustainable Development Agenda and Sustainable Development Goals (SDGs). 17 Goals to Transform Our World. http://www.un.org/sustainabledevelopment/. Accessed 23.04.2018

UN-HABITAT (2016) New Urban Agenda, adopted by HABITAT III conference, Quito, Equador. http://habitat3.org/wp-content/uploads/NUA-English.pdf. Accessed 23.04.2018

Valentin A, Spangenberg J (2000) A guide to community sustainability indicators. Environ Impact Assess Rev 20:381–392

Chapter 6
Determining Land Values by Location: Supporting Public Valuation Expert Committees in the Provision of Market Transparency

Andreas Ortner, Matthias Soot, and Alexandra Weitkamp

6.1 Location and Land Values in Development Processes

Developing land involves an entire process from planning through implementation. The aim is the subsequent use of land by the targeted group (the users). Concrete planning should be carried out for any sustainable implementation in the fields of economic and territorial development. In addition to the well-established (urban) land-use planning processes with their comprehensive standards and general lead function, newer informal strategic planning approaches are increasingly attracting attention. Here planning is implemented using a bundle of measures and concrete realisation steps. This affects several planning levels and fulfils the requirements of different planning horizons. Besides, handling current spatial planning challenges needs—in particular with regard to the increasing scarcity of financial and personal public resources—explicit spatial and temporal priorities have to be determined in order to realise development concepts or projects (ARL-net 2017). Planning and location are highly correlated to each other. Land-use planning has to take into account the existing locations (neighbourhoods). Otherwise, planning can change locations: each urban remediation measure tries to upgrade the quarter, each green-field development transforms arable land to building land. Each land-use change is also related to a change in land values. Modern land-use planning involves considering implementation issues—especially financial aspects. Land value increases caused by planning and implementation are an important key issue. The starting point and therefore the central issue of every project development is the value estimation (land and building) for the finished project (Alda and Hirschner 2014; Hendricks et al. 2017).

A. Ortner (✉) · M. Soot · A. Weitkamp
Chair of Land Management, TU Dresden, Dresden, Germany
e-mail: andreas.ortner@tu-dresden.de; matthias.soot@tu-dresden.de; alexandra.weitkamp@tu-dresden.de

© Springer International Publishing AG, part of Springer Nature 2019
M. Finka et al. (eds.), *The Role of Public Sector in Local Economic and Territorial Development*, EAI/Springer Innovations in Communication and Computing, https://doi.org/10.1007/978-3-319-93575-1_6

Implementation follows the planning. It can be divided into the initial preparation of the building land, e.g. by readjustment measures, and the subsequent concrete building code. According to the ideas of Agenda 21, local authorities are asked to pursue their own strategies for ensuring sustainable development on site (UN 1992). Besides, the multiplicity of the stakeholders involved plays an important role. The integration of diverse professional competences in the project development process diminishes the chance of unexpected challenges during project realisation. Moreover, development projects should have a high implementation orientation with a specific focus on the individual economic and territorial conditions of a certain location. In particular, this supports the long-term success of a project in all respects (Alda and Hirschner 2014; Gondring 2013; Klein et al. 2018).

The provision of market transparency is an essential task of the public sector. It guarantees the best possible conditions for the development and implementation of projects because all market participants need a variety of relevant data for purchase and sale decisions (Gudat and Voss 2009). Different professional groups like spatial planners, geographers, geodesist, engineers, economists, and property agents collect and process location and market data, as well as further information about the natural and the human-created environment. Land value is one of the most important indicators of location and is often used representatively. Based on the knowledge of public valuation expert committees, information for determining land values is highly dependent on purchasing price data. Nevertheless, project developers are aware that such data represents only part of the location as land values are also affected by other less influential parameters, or by parameters for which data is either not available or not at the correct scale. However, great difficulties are connected with the complexity of the available data (Kleiber et al. 2017; Weitkamp et al. 2017).

The provision of an advanced method of determining land values for all locations nationwide can improve market transparency. Providing a comprehensive decision-making basis for all market participants and the public sector will support a better management of spatial developments (Jeschke et al. 2016). The practical implication of the concept is shown using the case of Germany. Thereby, the municipalities are public authorities that work independently and that also function as both moderators and steering bodies between public and private interests. In general, the public experts fulfil important duties and responsibilities in the public interest and to benefit common welfare.

6.2 Relevance of Land Values for Spatial Developments

Project developments are important for urban development. The aims often comprise the realisation of innovative economic and territorial land uses. Accordingly, Alda and Hirschner (2014), project developments create real estate that must be in line with the market and whose use focuses on sustainable assets.

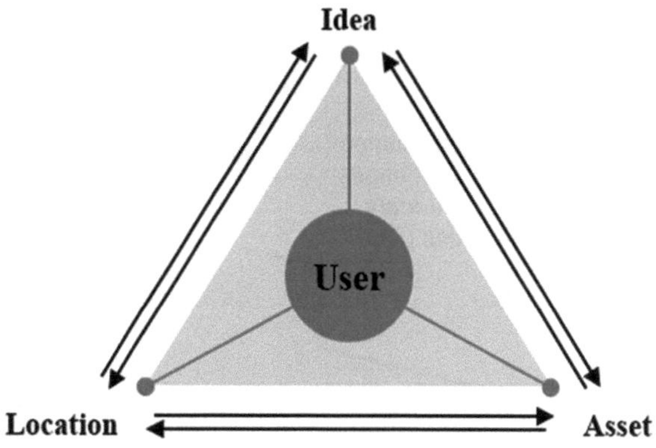

Fig. 6.1 Factors of the project development (based on Alda and Hirschner 2014)

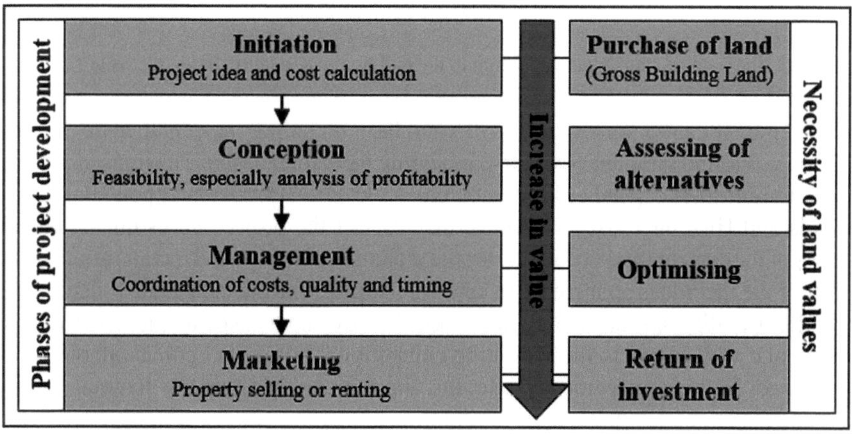

Fig. 6.2 Financing aspects in regard of relevance of land values in the project development process (based on Alda and Hirschner 2014)

A large variety of different factors play an important role in the project development process which can be directly related to innovative local economic and territorial advancements. Besides, project development has three main factors: idea, asset, and location (see Fig. 6.1).

"Classical" project development can be divided into four phases. These projects are always linked to a spatial component and realised on one or several properties.

Thus, land values play a major role in all development stages (see Fig. 6.2).

In the initial phase of a project, the purchase of land occurs—the developer buys the gross building land in expectation of earnable profits. These expectations can lead to land values that are frequently higher than those of arable land. Often, the area is actually presented as building land in preparatory land-use plans. The conceptual phase concentrates upon the feasibility study. In terms of costs, project

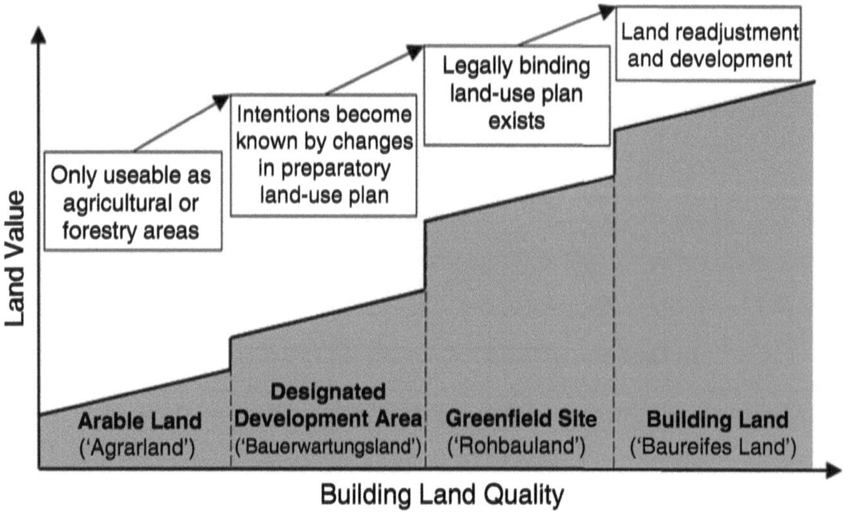

Fig. 6.3 Increase of land values according to the building land quality (based on Seele 1975)

development causes transaction costs (purchase including incidental costs), preparation costs, and building costs (also including incidental costs). The transaction and preparation costs depend especially on project location (particularly potential debts) (Alda and Hirschner 2014). Besides, the costs of the project are estimated on the basis of the current market conditions, the planning, and expected income afterwards. Part of the income is an increase in land value. Inter alia, the land value is dependent on the usability. On an underlying level, the developer takes a calculated risk in particular with regard to the prevailing competition. Often, an optimal alternative is prepared. In the management phase, the aim is that no external or internal factors should lead to higher costs, a longer duration or lower quality standards. Each delay or cost increase results in smaller gains. The better the management, the better the project will be. At the end of the project, marketing is undertaken. After selling or renting, the investment returns. The raised land values are part of this gain.

Location is the main factor which influences the value of a real estate. Therefore, micro- and macro-location as well as the property itself are crucial (Alda and Hirschner 2014). The more realistic the buildability becomes, the higher is the land value (see Fig. 6.3). This is valid for all sub-markets, no matter whether in the housing market or commercial sector.

In general, a rise in land value normally occurs through each development, even if there are differences in the initial level. There are substantial differences between developments on greenfield sites (with no building land quality and low land values) and restructuring measures affecting built-up land (building land qualities are existent but are affected by negative issues or lower urban density) (see Fig. 6.4).

Moreover, the location of the properties influences the land value. Developments often take place at the fringe of the settled area, while restructuring measures can be

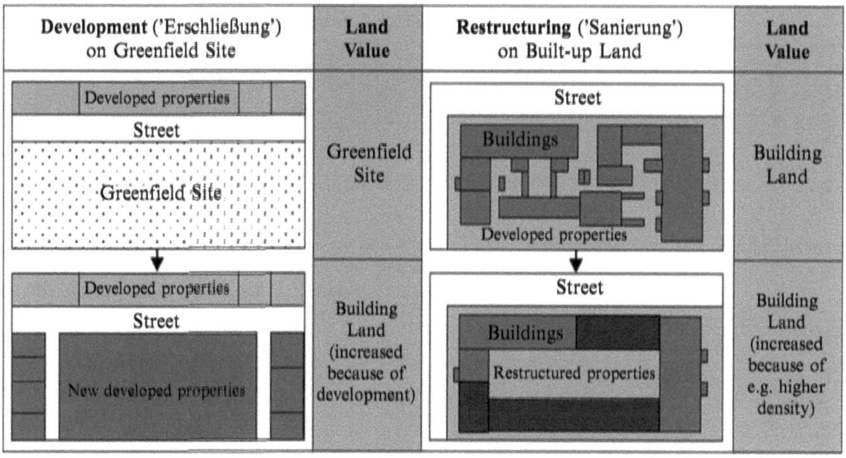

Fig. 6.4 Development versus restructuring of land (elaborated by the authors)

distributed throughout the whole settlement. Therefore, the determination of land values is highly relevant in urban and project development.

6.3 Determining Land Values for Valuating Locations

The land value is the fictional price for an unbuilt plot that a purchaser would pay in a free market under usual circumstances. It can be understood as an average of several purchases of the same plot in different buyer and seller constellations. Normally, the land value is given for a specific relative measure, e.g. Euros per sq. metre plot size. Furthermore, the real estate market is a dynamic market and prices change significantly over time (Gerardy et al. 2017). Thus, the land value is linked to a specific point in time. The prices that are paid for plots depend on three main factors (cf. Sect. 6.4 and Fig. 6.4): locational quality (e.g. neighbourhood), locational usability (e.g. residential or commercial purpose), and locational condition (e.g. shape, size, or slope) (Seele 1982).

The real estate market can be seen as being characterised by:

- Heterogeneity in offered goods
- Low supply elasticity to changes in demand due to the long production period of real estates
- Only partial market transparency
- A limited supplier and demander structure (but without the problem of monopoly)

Due to the location-bound and heterogeneous characteristics, spatial sub-markets exist. Each spatial sub-market can be divided into functional submarkets, e.g.

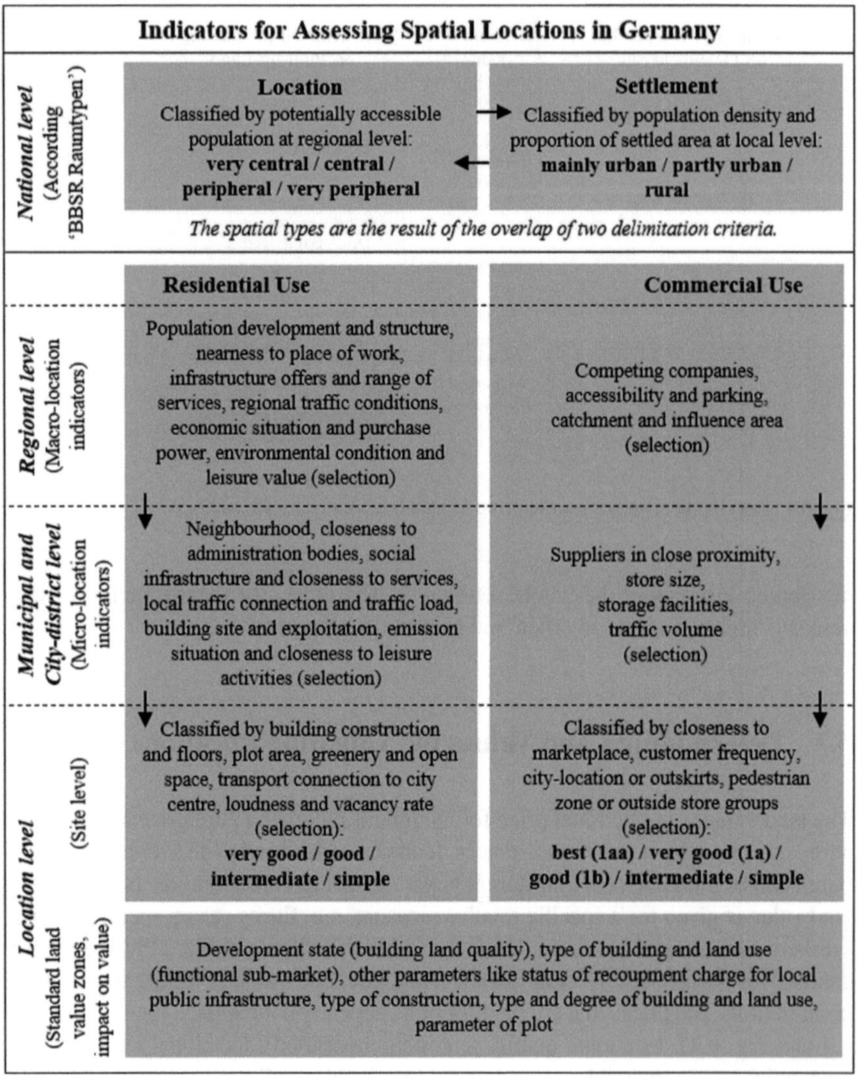

Fig. 6.5 Location indicators characterising locational quality (based on BBSR 2010; BMSBS 2010; BRW-RL 2011; Sprengnetter 1994)

unbuilt land, housing, social real estate, commercial. The functional sub-markets are affected by different spatial extensions (Brauer 2018).

The locational quality depends on a variety of location indicators (see Fig. 6.5). The example of Germany illustrates that location indicators are used on almost all planning levels. Nationwide, the Federal Institute for Research on Building, Urban Affairs and Spatial Development classifies different spatial types for planning approaches based on the overlap of two delimitation criteria: location (the potentially accessible population at regional level) and settlement (population density and proportion of settled area at local level).

These types characterise central and peripheral regions as well as urban and rural areas (BBSR 2010). Even at this subordinated stage, a (more or less qualitative) assessment of locations takes place as each person has their own interpretation of the meaning of urban or rural life and whether it is attractive to live within one of the spatial types. This already influences the demand for land within an area and therefore especially the level of land values.

The term macro-location refers to the country, region, and city level as well as the general supply, demand, and demand situation. The term micro-location encompasses the levels of municipality, city-district, and sub-centres (analogous to the macro-location). For simplification, in the following explanations macro-location indicators are assigned to the regional level and micro-location indicators to the municipal and city-district level in analogy to a spatial perspective.[1] The indicators can be distinguished between the different sub-markets, e.g. residential and commercial use. Accordingly, the indicators reflect rather personal preferences or economic interests. Nevertheless, the catalogue of indicators is not complete. Further criteria can be used to assess locations because most data exist at this level of consideration.

At the small-scale location level (properties and street), the assessment focus is on a concrete property. The quality of the construction, the plot area, the open space and greenery, and the transportation connections have an important influence on the classification into one of the four grade categories: very good, good, intermediate, or simple. Correspondingly, the land value estimated as high or low for residential use locations. Land values for commercial use are mainly dependent on the location's closeness to the marketplace and customer frequency. Categorised into best, very good, good, intermediate, or simple, land values decrease in line with quality.

The standard land values are derived from these quality assessments of locations. As already mentioned several times above, the public committees of valuation experts are responsible for the assessment of land values in Germany. The work of these experts has a far-reaching tradition. With the amendment of the building code in 1960 the price freeze on the real estate market from 1936 was abolished.

In this context, there was the need to observe the liberal real estate market to make it transparent for all market participants (Kleiber et al. 2017). To fulfil this task and to ensure market transparency, the legislators established the public committees of valuation experts. Besides deriving data for appraising real estate, one of their main tasks is to observe the market of undeveloped plots and derive the so-called standard land values (special feature in Germany).

In Section 2 of BRW-RL (2011) of the directive of standard land values, the standard land values are defined: *"Standard land value is the average land value of unbuilt land for a majority of plots within a demarcated area (standard land value zone). The plots are broadly consistent in terms of their land features, in particular the type and degree of usability, and are characterised by substantially equal*

[1] From a valuation perspective, the levels vary according to the spatial extent of the functional sub-market. For most common functional sub-markets, the macro-level contains the municipality and the micro-level includes neighbourhoods (or better standard land value zones).

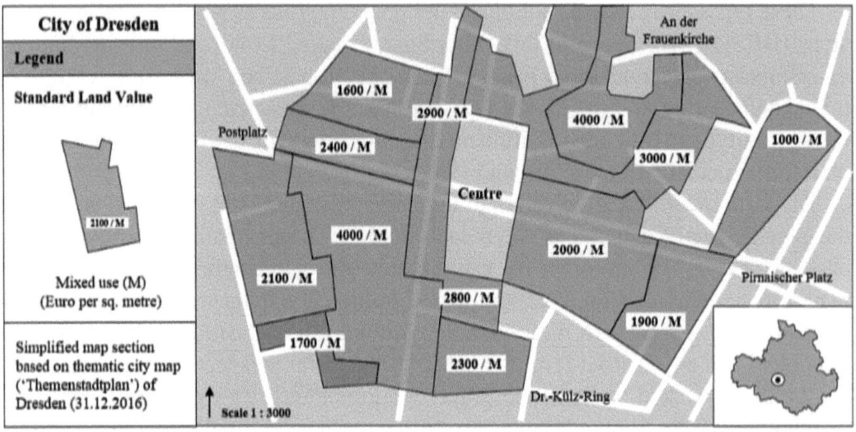

Fig. 6.6 Standard land values for mixed use (Euro per sq. metre) in Dresden (Germany) in 2017 (simplified map section based on thematic city map ("Themenstadtplan") of Dresden)

general value ratios. The standard land value is based on a square metre plot of land with the property features indicated (standard land value plot)".[2]

The standard land values are timed values for a specific (fictional) unbuilt plot in a defined homogeneous (structure and value) zone (see Fig. 6.6). The plot indicated is defined in its characteristics (usability and condition). Since the amendment of the valuation law (ImmoWertV) in Germany in 2010, standard land values must be derived comprehensively and published nationwide. The regular evaluation period is at least every 2 years. Standard land values enable the provision of nationwide information about the value of land for a specific date for planners and developers. Economic developments on specific markets can be investigated using these values, as can the general price level in specific markets and locations.

Standard land values are not a regulatory instrument. They are intended just to describe the market and help to make it transparent (Ernst et al. 2014). The observation of these values gives an idea of economic developments and changes in locational quality. Thus, the standard land value can be used as an indicator for location in valuations (Weitkamp et al. 2017).

To ascertain comprehensive standard land values, first land values must be determined. The general approach to determine the value of a plot is the comparison approach. This is the closest to the market of all methods for deriving land values. First, the available prices of undeveloped plots which are sold in the same region, at a specific time and which have the same usability should be collected. At least 5–15

[2]In German: *"Der Bodenrichtwert ist der durchschnittliche Lagewert des Bodens für eine Mehrheit von Grundstücken innerhalb eines abgegrenzten Gebiets (Bodenrichtwertzone), die nach ihren Grundstücksmerkmalen, insbesondere nach Art und Maß der Nutzbarkeit weitgehend übereinstimmen und für die im Wesentlichen gleiche allgemeine Wertverhältnisse vorliegen. Er ist bezogen auf den Quadratmeter Grundstücksfläche eines Grundstücks mit den dargestellten Grundstücksmerkmalen (Bodenrichtwertgrundstück)"* (BRW-RL 2011).

purchase prices are required for this approach (depending on the similarity and scattering of prices). If there is a sufficient number of purchases, the differences in quality of each plot should be converted with the (local) available coefficients. In most cases, only coefficients to convert the condition of the plot are available (Mann 2014). If past prices are available, the economic development on the specific market of undeveloped plots should be adjusted. The committee of valuation experts has to derive such indexes for the relevant markets.

6.4 Supporting Market Transparency

Market transparency is not equal everywhere. In some areas and regions, only a few prices are available. Soot et al. (2017) developed a system to identify different regions with a lack in the number of transactions (see Fig. 6.7). Especially a difference between rural areas and city centres could be identified. If there are no or only few transactions (purchases) available in the specific zone, other comparable regions (same regional quality, usability, and condition) should be considered to find comparable sold purchases. To identify comparable regions, the locational quality should be investigated. The above-mentioned factors that influence locational quality can be used to identify matching areas for the search for other purchases in the market.

As mentioned, the general small number of purchases leads to the problem that land value is difficult to derive (Weitkamp et al. 2017). In a survey, Reuter (2006) found various methods that are used in practice if there are too few purchases. The main methods are deductive methods and intersubjective methods. Deductive methods use other data to gather the price of the undeveloped plot, while intersubjective methods use the expertise of professionals.

The guideline for standard land values differentiates between three common valuation types: costs and rents methods (both are deductive methods) and intersubjective methods.

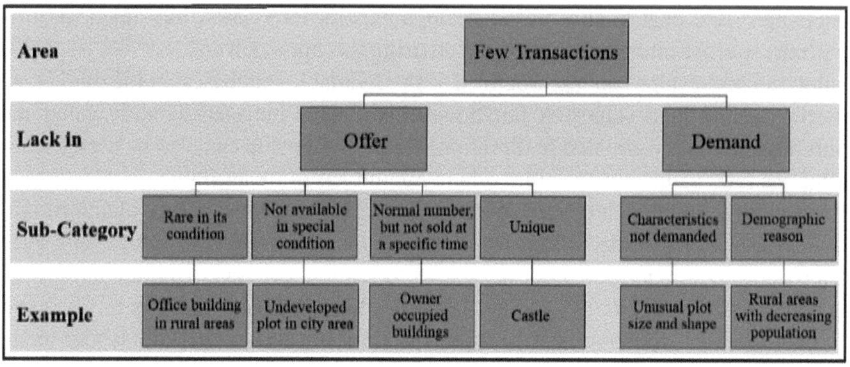

Fig. 6.7 Areas with few transactions (according Soot et al. 2017)

Costs method The trading of plots in the stages of designated development areas and Greenfield sites (see Fig. 6.3) is rare and only very few transactions can be found in this market segment. To derive the value of such plots, Seele (1998) presented an approach for estimating the value of untapped building land (UBL) from building land. He suggests reducing the price for building land (BL) by the costs for developing (*C*), considering the proportion of land needed for infrastructure (*f*), and discounting it with waiting time (*n*). The rate (*p*) for the discount should be derived from the market. Alternatively, it is suggested that the rates from the income method in the specific market be used (see Eq. 6.1).

$$\mathrm{UBL} = (\mathrm{BL} - C) \cdot \left(1 - \frac{f}{100}\right) \cdot \left(\frac{1}{q^n}\right) \quad \text{with} \quad q = 1 + \frac{p}{100} \tag{6.1}$$

Rents method Rents are frequently used as data for deductive methods. Different approaches can be found all over Germany. Fieder (2006) used rents from the first-floor stores in the city centre of Dresden to derive the standard land values. Fiedler derived a function between purchase prices and rents and thus predicted the value of areas with no transactions using the known durational rents in the specific zone. Alternatively, Strotkamp (2006) combined the rents of stores and residential uses to derive this relation.

Intersubjective method (experts based) Another way to assess land value is the so-called intersubjective method. Here, a number of experts are interviewed. The experts should have the same level of information and try to objectify their statements. The genesis of answers should be reproducible and transparent (calculation and justification should be attached for each statement). Sometimes a Delphi method is applied to homogenise the results. Examples of such methods can be found in (Hagedorn 1984; Hildebrandt 1976; Jeschke 2011; Strotkamp 2012).

A common method to objectify the estimation of the spatial quality involves the use of an objective-tree method (e.g. cf. Jeschke (2011). Experts are asked to access locational quality within several categories that influence the locational quality in the specific market. The different categories are weighted (which category influences the locational quality more) through experts interviews, too. Each location within a specific market can be estimated using this approach and receives a specific value (location value) on a pre-defined scale. Available purchases on the market are used to derive land values. A functional relationship between location value and purchase prices is estimated (calibration). This relationship can also be used to estimate the land values for locations where no purchases are available.

Deductive and intersubjective methods are widely accepted in the valuation community and by the jurisdiction. However, the derivation of land values with such approaches is a laborious process. Access to rents is generally limited and indeed impossible in a lot of municipalities because of strict privacy policies in Germany. Expert interviews (intersubjective approaches) for big sub-markets are labour intensive and have to be redone regularly to identify changes in the quality (recalibration

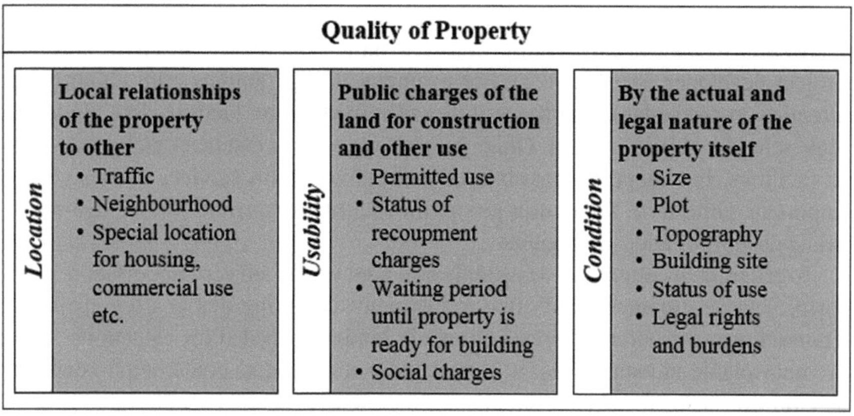

Fig. 6.8 Influence on the quality of property (elaborated by the authors)

of the model). For a reliable expert-driven estimation of locational quality, a sufficient number of experts is needed. Jeschke et al. (2014) used five experts for estimation in only one housing sub-market of a smaller city in Saxony.

Further value-relevant factors should be included in the valuation process to support a more reliable result—especially in areas with little information (lack of purchases). Thus, there is currently an increasing focus on other value-relevant factors in the hope of being able to carry out an even more exact assessment (see Fig. 6.8). Therefore, cross connections to criteria like location factors or accessibility should be more strongly acknowledged (Weitkamp et al. 2017).

The location of a property is defined by its relationships to other properties. Parameters like traffic situation and neighbourhood are relevant. The usability is mainly determined by the public charges and planning circumstances. For example, the value of plots that are usable for residential purposes differs from plots which are usable for commercial purposes in same area. In general, plots in rural areas are cheaper than in city centres due to usability. Again, the location is the main component which influences the price within one market segment of a plot. The third criteria of property quality is the condition in terms of its actual and legal nature. Characteristics like plot, size, topography, building site, or status of use are value relevant.

For project planning and development, the "softer" characteristics of a location like the socio-cultural (public) image (e.g. whether a neighbourhood is attractive or not) in relation to different socio-economic user groups in a location (e.g. city-district) also matter in the decision-making process. As a result, they also have a relevant effect on value. Even if the users of a plot (e.g. tenants of an apartment or house) do not typically correspond with the purchaser of a plot, their perceptions and needs are of relevance. They affect the return on investment by paying the rent. The higher the attractiveness, the higher the income is.

Insofar, the success of a project is dependent on whether it is accepted after realisation in the certain location. This has indirect influence on the profit of the project developer (see Fig. 6.2). For example, young families with children pay attention in particular to the accessibility of infrastructure facilities like kindergartens, schools, or playgrounds. Older people expect short distances and good access to facilities, but shopping possibilities and public health services are also rather important. Otherwise, for all user groups the security (e.g. crime control, age-appropriate design) of a location counts.

To reiterate, location and, as its indicator, land values are very important not only for project development. But often in Germany, capturing this is a big challenge. Transactions (the purchase prices of unbuilt land) are needed for valuations. These are unavailable in many areas. It is expected that other data could help to compensate for this lack. However, it remains the case that much data is not available at the necessary scale (resolution of properties or neighbourhoods, like standard land value zones).

The attempt must be undertaken to utilise existent information and data on larger scales, also for the small-scale location level. It can be stated that new requirements for the collection, calculation, and management of location data exist (Ortner and Weitkamp 2018). For the formulation of guaranteed statements (which indicate the status quo of a location), it is also important that the overall dynamics of change are not too high. The modelling of temporal influences is difficult to handle. Research on well-established and new approaches to determining land values is of high relevance, as are the current methodological limitations, as the findings can support the public valuation expert committees in the provision of high market transparency. In addition, the determination of land values is one of the most important indicators which affects development decisions.

6.5 Conclusions

Land values play a crucial role in the process of project planning and development. Comprehensive statements on land value—determined by the characteristics of locations—support the provision of market transparency. In Germany, an established assessment system for land values has been set up over many years. It allows objective and secure statements about land values to be made. The determination of standard land value zones for different location usability provides nationwide comparability of locations.

However, sufficient information about prices or location factors are not available in all areas and regions of Germany. Here, the public valuation expert committees use various methods to generate reliable statements about land values. Furthermore, the catalogue of indicators is not complete. Other value-relevant factors should be used to assess locations. Accessibility by public or private transport systems, public image, and the demands of purchasers and user groups, for instance in terms of the availability and security of infrastructure facilities, are gaining significance.

Therefore, scientists and the public valuation experts try to cooperate closely to compile new possibilities to complement established tools with appropriate new methods and instruments to increase the basis of information on locations and to raise market transparency.

The German experiences can be helpful for other European countries. The supply of objective information and—due to the public valuation experts knowledge—the market transparency is high in Germany. The establishment of nationwide working instruments strengthens the planning and implementation of development projects and offers more safety and reliability for all involved stakeholders.

References

Alda W, Hirschner J (2014) Projektentwicklung in der Immobilienwirtschaft. Springer Fachmedien, Wiesbaden

ARL-net (2017) Strategic planning. https://www.arl-net.de/lexica/de/strategische-planung?lang=en. Accessed 10 Jan 2018

BBSR (2010) Laufende Raumbeobachtung—Raumabgrenzungen. Raumtypen 2010. http://www.bbsr.bund.de/BBSR/DE/Raumbeobachtung/Raumabgrenzungen/Raumtypen2010_vbg/Raumtypen2010_alt.html. Accessed 5 Jan 2018

BMSBS (2010) Verordnung über die Grundsätze für die Ermittlung der Verkehrswerte von Grundstücken (Immobilienwertermittlungsverordnung—ImmoWertV) V. v. 19.05.2010 BGBl. I S. 639 (Nr. 25). Geltung ab 01.07.2010

Brauer KU (2018) Grundlagen der Immobilienwirtschaft. Springer Fachmedien

BRW-RL (2011) Richtlinie zur Ermittlung von Bodenrichtwerten (Bodenrichtwertrichtlinie—BRW-RL) vom 11. Januar 2011. BAnz Nr 24 vom 11/02/02011, p 597

Ernst W, Zinkahn W, Bielenberg W, Krautzberger M (2014) Baugesetzbuch: Kommentar. Beck-Online, München

Fieder P (2006) Zur Nutzung von Geschäftsmieten für die Überprüfung und Verdichtung von Bodenrichtwerten in der Landeshauptstadt Dresden. Diplomica Verlag

Gerardy T, Möckel R, Troff H, Bischoff B (2017) Praxis der Grundstücksbewertung (Loseblattsammlung). Olzog, München

Gondring H (2013) Immobilienwirtschaft: Handbuch für Studium und Praxis. Franz Vahlen, München

Gudat R, Voss W (2009) Transparenz am Grundstücks- und Immobilienmarkt: Verlässlichkeit behördlicher und gewerbsmässiger Marktinformationen. Flächenmanagement und Bodenordnung (1):19–33

Hagedorn W (1984) Ausgleichsbeträge nach Standortverbesserungen im Sanierungsgebiet. Städte- und Gemeindebund, pp 624–628

Hendricks A, Kalbro T, Llorente M, Vilmin T, Weitkamp A (2017) Land ownership and land use development—the integration of past, present, and future in spatial planning and land management policies, vdf Hochschulverlag, Zürich, chap Public Value Capture of Increasing Property Values—What are "Unearned Increments"? pp 257–282

Hildebrandt H (1976) Bewertung in der Sanierung mit Hilfe relativer Lagewerte. Vermessungswesen und Raumordnung 38(2):65–70

Jeschke A (2011) Flächenhafte Ermittlung von Bodenrichtwerten mittels Lagewertverfahren. Flächenmanagement und Bodenordnung (FuB) 1(73):25–31

Jeschke A, Krägenbring R, Steinbrecher D (2014) Zur Optimierung der Auswertung intersubjektiver Schätzungen—Psychologische, mathematisch-statistische und messtechnische Aspekte bei der Auswertung von Expertenbefragungen. Flächenmanagement und Bodenordnung (FuB) (2):83–89

Jeschke A, Weitkamp A, Köhler T (2016) Stakeholders in land management processes—an approach of game theoretical modelling. avn Allgemeine Vermessungsnachrichten (2):39–49

Kleiber W, Fischer R, Werling U (2017) Verkehrswertermittlung von Grundstücken. Bundesanzeiger Verlag, Köln

Klein IM, Jahn A, Weitkamp A (2018) Derelict Land – A Chance for Affordable Housing in Germany? In: EALD-Proceedings (accepted)

Mann W (2014) Marktransparenz durch Boden- und Immobilienrichtwerte. In: Das deutsche Vermessungs- und Geoinformationswesen: Themenschwerpunkt 2014: Stadtentwicklung und Städtebau, Stadterneuerung und Stadtumbau, Flächenmanagement und Bodenordnung, Immobilienmärkte und Grundstückswertermittlung; Handlungsfelder und Struktur, Berlin, pp 581–614

Ortner A, Weitkamp A (2018) "Räumliche Lage" – A Discourse on Location – Concepts for Assessing Interdisciplinary Application Potentials. In: EALD-Proceedings (accepted)

Reuter F (2006) Zur Ermittlung von Bodenwerten in kaufpreisarmen Lagen. Flächenmanagement und Bodenordnung (FuB) 3:97–107

Seele W (1975) Bauleitplanung und kommunale Bodenpolitik als Determinanten der Baulandpreisentwicklung. Vermessungswesen und Raumordnung 37(3):138–154

Seele W (1982) Wertermittlung bei der Preisprüfung und Ungewißheit des Verkehrswertes. Vermessungswesen und Raumordnung 44(3):105–122

Seele W (1998) Bodenwertermittlung durch deduktiven Preisvergleich. Vermessungswesen und Raumordnung 98:393–412

Soot M, Weitkamp A, Dorndorf A, Alkhatib H, Jeschke A (2017) Different regions with few transactions—an approach of systematization. In: FIG Working Week 2017, Helsinki

Sprengnetter HO (1994) Immobilienbewertung (Loseblattsammlung). Sprengnetter, Sinzig

Strotkamp HP (2006) Ermittlung von Bodenwertänderungen aufgrund von Lage- und/oder GFZ-Unterschieden mittels Mietsäulenverfahren. Flächenmanagement und Bodenordnung (3)

Strotkamp HP (2012) Ermittlung von Boden(richt)werten mittels Zielbaummethode—Zur Renaissance der Zielbaummethode. Zeitschrift für Liegenschaftswesen, Planung und Vermessung (1):2–9

UN (1992) Agenda 21. In: United Nations Conference on Environment and Development, Rio de Janeiro

Weitkamp A, Koehler T, Ortner A (2017) Bodenwertermittlung—Eine automatisierbare Aufgabe? Flächenmanagement und Bodenordnung 78(1):25–34

Chapter 7
Forest Commons as a Model for Territorial Governance

Rudina Toto

7.1 Introduction

In this chapter, the discussion is made on the growing importance of the concept of territorial governance as the new phase of good governance, which is polycentric, multi-level, multi-actor, multi-purpose, and multi-sector. This is the governance of elusive, fuzzy, and soft spaces/territories (Finka et al. 2015) and of resilient institutions and resources. The illustration of this discussion is made through the case of forest commons governance in Albania. The Albanian commons' regime is informal and thereby thought to be very weak, but surprisingly present and active. The way the system has survived endogenous and exogenous factors of socio-political changes, is by employing dimensions and features of territorial governance, and overcoming some of the gaps that environmental governance has to deal with in the very complex context of human–environmental relationships, namely: the human society value reference system, the lack of ecosystem principles in conventional governance, and scale dependencies or mismatches among local solutions and global challenges.

Unravelling the theoretical discourse and by using the case of Albanian forest commons, the chapter explains how territorial governance: (1) has all of the properties that a governance system requires for enabling sustainable development; (2) is embedded within and across sectors and scales, therefore being the outcome of multi-sectorial and multi-stakeholder performance, rather than merely of government institutions; (3) can very well provide a path or a model for scientific knowledge transfer from experts to lay people and vice versa, for traditional knowledge from community institutions to experts, as the means for bridging the gap between

R. Toto (✉)
Permanent Workshop on Territorial Governance, Co-PLAN, Institute for Habitat Development, POLIS University, Tirana, Albania
e-mail: rudina_toto@co-plan.org

© Springer International Publishing AG, part of Springer Nature 2019
M. Finka et al. (eds.), *The Role of Public Sector in Local Economic and Territorial Development*, EAI/Springer Innovations in Communication and Computing, https://doi.org/10.1007/978-3-319-93575-1_7

small-scale governance and successful solutions on one side, and large-scale conventional governance and global challenges on the other.

An attempt is made to converge the forest commons governance (territorial, traditional, and local scale) with ecosystem services-based conventional governance (territorial, scientific, multiple-scales), by first identifying where the divergences may be, and then proposing a model for convergence. Again the model is illustrated with findings from the case of Albania's forest commons. The case identifies the possibility for convergence and its expected positive outcomes, as well as signals the needs for future research and work in the territorial governance and development domain.

Albania, as a case, is a country that in the early beginning of 1990s went through a dramatic political and economic system shift, from a centralised economy to the market one. Major changes in terms of social aspects, territory and resources, governance and citizens' mentality were the immediate outcome of the political turmoil. Thousands of people migrated from the remote rural areas to the urban centres in search for better services and economic development opportunities. This all resulted in a growing need for affordable housing. As the government could not respond to this high pressing demand, the newcomers crafted their own individual solutions, characterised by occupation and fragmentation of agriculture land, water pollution, informal deforestation, wetlands pollution and alteration, and losses on biodiversity including species' extinction risk. This behaviour was initially considered as "a natural response" to immediate needs. However, neither the community, nor the institutions thought about the implications and costs born with the loss of the ecosystem functions (Shutina and Toto 2010).

Despite progresses made towards improving governance so as to achieve sustainability, problems like monocentric development, compromised ecosystems and natural resources, low economic and ecological resilience remain strongly present. In order to foster territorial development, through innovative and knowledge transfer means, governance in Albania has to reinvent itself towards embracing the territorial perspective. The latter is dependent on the social capital and socio-ecological interactions. The high emigration rate constitutes a wakening factor for the domestic social capital. Furthermore, rural-urban migration leads towards lost connections between ecosystems and users. This leaves ecosystems unprotected and their services heavily exploited by others who have no traditional and historical ties, and therefore, in absence of the proximity principle, places a very high discount rate on the resources.

Methodologically speaking, the model proposed in this chapter is based on bringing together current work made in the studies of territorial governance, socio-ecological systems, common pool resources, and ecosystem services. The model of forest commons is explored in the Albanian case of forest commons through the lenses of the CPR theory, and then analysed for its larger scale by adding more variables. The territorial analysis of these variables and of the ecosystem services provides food for thought to the assessment of the territorial governance dimensions for forest commons. Evidences are collected through field visual surveys, interviews and focus groups, distributed over the territory for a complete representation of all common forests and systematically researching a number of common pool resources' variables of endurance and functionality.

7.2 Territorial Governance

Governance Shifts and the Territory

The increasing complexity of the society and its radical changing implies also shifts in governance (Kovač et al. 2016; Driessen et al. 2012; Arnouts et al. 2012) that are associated with new knowledge on the notion and on the related practices. Governance—denoting an entirety of policy decision-making processes, stakeholders, interactions, and legitimacy (Rhodes 1996; Stoker 1998; Driessen et al. 2012) represents both, the way people chose to advance development and the outcome of societal transformation (Rhodes 1996). Therefore, the modes of governance and their shifts indicate also a status of socio-institutional evolution and thinking (Rametsteiner 2009), and are challenged by the "emergence of new knowledge dynamics" (Halkier 2012, p. 1767). Various authors analyse governance shifts, focusing on the critical features and dimensions of the governance modes, their policy, and concrete effects and outcomes, as well as on the supporting theories of public policy and administration. Araral and Hartley (2013) explore the paradigms of public administration, laying emphasis on polycentric governance as the most recent mode, governance is heading towards on a paradigmatic level. In their analysis of public administration generations, they transit from traditional management, to new public management and further on to collaborative governance (Araral and Hartley 2013). While the latest is the recurring mode, its successes and failures lead towards the need for polycentric governance.

Early research on polycentric governance dates as of the 1950s and has a peak in 1961 with Vincent Ostrom et al. Writing their "theoretical inquiry" on the "organization of government in metropolitan areas" (p. 831). They define the term "polycentric" as an entirety of decision-making centres, formally independent from one another. If these centres carry out functional interaction and do so to a level of establishing interdependent relationships, then this entirety becomes a system (Ostrom et al. 1961). This definition of polycentric governance is important in several aspects:

1. First, formulated in the 1960s, it very well precedes the discourse on the conceptual and practical transition from government to governance. This discourse emerges 20 years later on the assumption that the role of the government on socio-economic affairs should be reduced so as to avoid unique prevalence (Alcántara 1998) and remains present to date. In his analysis of "from government to governance", Hysing (2009) refers to Hajer and Wagenaar (2003) and Kooiman (1993) for defining the state as unable to pursue collective action—The latter being the role of governance (Olson 1965/2002) because of the increasing complexity of governing challenges. These challenges relate, among others, to the growing number of stakeholders and their [often conflicting] needs, and to the controversial process of natural resources exploitation that impacts the welfare, and for that matter also the stability of the stakeholders' groups.

Arnouts et al. (2012) Present a typological framework for operationalizing governance modes and shifts, based on Kooiman (2003), with four modes in a continuum: Hierarchical, closed, and open co-governance and self-governance. Driessen et al. (2012) Build as well a conceptual framework for differentiating between modes of environmental governance, starting from the assumption that environmental governance is the "answer to degrading environmental quality, depletion of the resources, biodiversity loss and climate change" (p. 144), being therefore a critical mean to achieve sustainable development (Toto 2018; Meadowcroft et al. 2012). Actors and powers are key denominators of the analytical matrixes in both cases. Therefore in one pole stands the government as the dominating actor, who allocates power to other stakeholders based on chosen levels of balance between centralization, decentralisation, and collaboration (Driessen et al. 2012; Arnouts et al. 2012). On the other, there is governance with non-government actors, who have unaffected autonomy, are free to initiate processes based on their willingness, and based on power that is inherent and non-allocated to them (ibid.). One fundamental conclusion of both studies rests with modes of governance not fully replacing, but building on one another and more empirical research being needed to understand the ability of both, transitions between and co-existence of governance modes to spur successful societal change for sustainability. This conclusion is critical to the following analysis on territorial governance.

2. Second, it places an emphasis on the concepts "multi" and "interaction". So, the several decision-making centres could be multi-actor/sector/level/issue-oriented, or could as well be a combination between two, or more of these. The interaction evolves from interdependencies of these various combinations and, based on the distribution of power, knowledge, and ability for self-decision making, reveals the modes of making governance and the transformations the latter bears with time. After all, this interdependent engagement is the essence of governance [Treib et al. (2005) in (Arnouts et al. 2012)] and, because it is dynamic and in a continuously changing state of equilibrium (Driessen et al. 2012), it contributes to the argument of governance remaining still an elusive concept (Arnouts et al. 2012), regardless of the efforts for establishing governance theories. According to McGinnis (1999) in Araral and Hartley (2013), the ability of groups to engage in self-governed actions and decisions is fundamental to the concept of governance. Ostrom (1990, 2009b) Is adding to this capacity also three other features, i.E. fit-to-context and adaptability of these self-governed institutions as well as the proper scale of management, often translated as the governance principle of subsidiarity.
3. Finally, the definition of Ostrom et al. (1961) contributes also to further studies on polycentrism, such as those related to territorial development, territorial multi-level governance, and ecosystem-based governance. Of special interest has been the discourse on European Union multi-level governance, but also on what Hooghe and Marks (2003) describe as type II multi-level governance—Fragmented into functionally specific issues, jurisdictions, or diverse and flexible scales of interaction. Ecosystem-based governance is also uprising, with the

understanding that network and borderless governance is appropriate for dealing with environmental concerns and the resiliency of ecosystem services because of the latter not fitting into jurisdictional boundaries.

Araral and Hartley (2013) argue that polycentric governance is "relevant to the discourse on post-governance" because "modern societies embody institutional diversity, reflected in multi-level, multi-purpose, multi-sectorial, and multi-functional units of governance" (p. 2). A way to deal with this diversity is by employing multi-level governance, a notion that focuses on various policy tiers attributed to different actors (Arnouts et al. 2012) and is quite flexible because of not concentrating governance in one single level or jurisdiction that would otherwise hamper the efficiency of managing and delivering public goods (Hooghe and Marks 2003; Finka and Kluvánková-Oravská 2010). Multi-level governance emerged as a concept in European Union (EU) studies in the 1980s, but there has been ambiguity between the conception as an EU multi-level polity on the one hand and as a reflection of the transition from government to governance on the other (Faludi 2012). Faludi argues by referring to Piattoni (2010) and Healey (2006) that the ambiguity stems from the confusion between networks of organisational interaction established within and among government institutions, and multi-purpose networks of collective action. The latter is in fact reflected within the good governance principles of engagement, subsidiarity, equity of access to decision-making, and transparency and accountability (UN-Habitat 2002) that go beyond what Faludi (2012) describes as "vertical relations between bodies of government within a multi-level polity" (p. 207)—the original meaning of multi-level governance from an EU perspective.

While the literature and theoretical discourse on multi-level governance has advanced in the recent years to examine its characteristics and shape the concept, territory as a conception is limitedly incorporated in the research (Well and Schmitt 2015). Yet, "territory matters in governance, being its ultimate object as a reflection of socio-ecological interactions and interests" (Shutina and Toto 2017) that governance has to deal with. In their analysis of territorial governance, Davoudi et al. (2008) lead towards the understanding of the links between the territory and socio-ecological interactions (though with no explicit use of these terms) in the broader frame of governance aims. Thus, they define governance as the capability of actors to engage and foster collective action through a consensually strategizing process on the future visions for the territory. Furthermore, they define territorial governance as governance that is able to develop resilient territorial capital (OECD 2001; Davoudi et al. 2008), where the latter, as Davoudi et al. (2008) argue, represents the endogenous potential of the territory, reflected in its natural and manmade resources, commonly owned and managed, and place specific. The interactions between stakeholders, as well as their policy decision-making reflect the given character of the territory where they perform, while the outcome of their interaction affects immediately the territorial capital. Therefore, analysing territorial governance in the view of Davoudi et al. (2008) should consider both, the context and the interactions, with the second including territorial governance actors, policies, and actions.

In his criticism for multi-level [territorial] governance, Faludi (2012) focuses also on the "territory" discourse, trying to distinguish between the territory as a "fixed container" (p. 204) with jurisdictional boundaries of government, and the territory as a soft space of fuzzy boundaries, and of flows and interactions (Finka et al. 2015; Schmitt et al. 2013; Faludi 2012; Raymond 2008; Blatter 2004). The latter conception on territory relates to its definition as a socio-political construct (Davoudi et al. 2008; Schmitt et al. 2013), therefore on territory as a representation of interactions and governance. This corresponds with the "relational space" concept of the territory that Faludi (2012) provides, where governance actions, for instance, planning, transcend jurisdictional boundaries. Wrecking "physical" boundaries in favour of elusive ones increases the difficulty to recognise the space and/or the scale, where a governance action is to be performed. It actually leads towards spatial overlaps that diffuse the territorial boundaries of the responsibility (Faludi 2012) and government's accountability, as well as have an impact (at least implied) over the voters' legitimacy.

Still, considering that the scale of economic units and urban territories does not match that of the territorial and natural capital, even less the scale of natural capital services, there is clearly a need to plan and manage out of the "territorial trap" (Faludi 2012, p. 208), by exploring alternative, soft territorial spaces and arrangements. This sounds very challenging, but it is possible to accomplish this difficult task by implementing governance actions through issue-based and place-based approaches. There is yet a risk for loosing or weakening the comprehensive view over the territory, or the strategic character of an objective (for instance, in strategic planning), due to possible fragmentation of the examination and decision-making process. There is also the risk that new institutions are created ad hoc and replace current ones without due thinking. Even so, focusing on the capacitation of current institutions to develop overlapping, but sound views/analysis of network space and problems, and especially galvanising hidden institutions by enabling knowledge transfer between levels and across stakeholders, are good mechanisms to overcome the risk. Overall, understanding and acknowledging that governance has to be studied with due attention to the territorial dimension is already an important achievement that would actually increase the social awareness on territory's diversity and hence on its impacts on policy decision-making (Schmitt et al. 2013).

Though recent in governance studies, territorial governance is bound by an evolving theoretical and policy discussion. Regardless of the criticism made to it, especially on the possible redundancy of using the term territorial next to multilevel governance (Faludi 2012), there is reason to assert that alike governance is a term in evolution since its separation from the "old" government (Stead 2014; Kovač et al. 2016; Driessen et al. 2012), territorial governance represents the next knowledge-making milestone of governance as a developing concept. The first definition of territorial governance comes from OECD in 2001: "the manner in which territories of a national state are administered … with particular reference to the distribution of roles and responsibilities among the different levels of government … and the underlying processes of negotiation and consensus building" (p. 142). Being a definition for the governance of the territories instead of governance with the territory as an inherent critical dimension, it coincides with regular governance, or at its best with multi-level governance in its original meaning. Initially, European Union institutions deployed this same definition of territorial governance, as it was the case for CEMAT Resolution on Territorial Governance (2006) (Stead 2014); and the EU Territorial Agenda 2007 (Well and Schmitt 2015). The second did not bring anything new to the concept, but the previous added the dimension of understanding territorial governance as the "emergence and the implementation of innovative shared forms of planning and managing of socio-spatial dynamics" (CEMAT 2010, p. 269), hence challenging the traditional role of state in spatial development, through evoking a multi-actors engagement approach.

Further policy processes occurring at the EU level, slowly acknowledged territory as a perspective in governance. Public debate on territorial cohesion in 2009 articulated the necessity for territorial dimension to "be reinforced at all levels and at all stages in policy design and implementation", reflected in the Green Paper on Territorial Cohesion—Turning territorial diversity into strength (European Commission 2009). Böhme et al. (2015) analyse a number of strategic documents such as Territorial Agenda 2020, Europe 2020, Cohesion-Policy 2014–2020, and funding mechanisms (particularly CLLD[1] and ITI,[2] but also the ERDF[3]), which encompass diverse levels of territorial sensitivity. Europe 2020, for instance, follows the subsidiarity principle, but is weakly understood by the citizens and has a low involvement of local governments in implementation, therefore being "territorially blind" (CoR 2014). It is for this reason that the Athens Declaration of the Committee of the Regions introduces the need to give to the strategy a territorial dimension as the first in a 7-points plan (CoR 2014). The Cohesion Policy on the other hand has a stronger territorial dimension, included also in the above funding mechanisms. It builds on territorial governance through shared management, partnership principle, and territorial tools. However, at strategic European level the territorial dimension seems becoming weaker with the funding shifting from regional to national programmes (Böhme et al. 2015). There seems to be more hope though in the local and regional level, where policy-makers are actually adopting the place-based approach (ibid.).

Territorial governance is not a directive at EU level, but it appears in the Cohesion objective (Rivolin 2010) and can be tracked in several sectorial policies, to mention transport, energy, environment, etc. A very first document that raises the importance of territorial governance at EU level is the European Spatial Development Perspective (1999). ESDP requests for coordination between the territorial development—the objective of territorial governance, and sectors, as expressed in its objectives of transportation, competitiveness, cultural, and natural heritage (CSD 1999), which are still valid to date. On the other hand, regardless of the extent in which territorial governance formulations are present in EU strategic documents, the success of their implementation depends on the territorial perspective that mem-

[1] Community-Led Local Development

[2] Integrated Territorial Investments

[3] European Regional Development Fund

ber and/or aspiring states employ in the process. For instance, the enlargement perspective of the EU for the Western Balkans provides for the reform actions that the region must undertake to achieve integration. Territorial governance formulations are not obvious in the text of the communication, but territorial sensitivity is strongly present. The communication stresses the imperative for achieving good governance and public administration reform to support governance shifts, strengthening socio-economic development through—among others—the establishment of a regional economic area, increasing connectivity with a focus on transport and energy (interaction and natural resources), and strengthening regional cooperation (European Commission 2018). The communication requires for proper balance between levels and scales of governance, increased accountability of the administration, better finances and improved public services and improved civil society (European Commission 2018), hence ingredients to achieve collective action.

Territorial Governance Dimensions

It is clear so far that territorial governance is a milestone in governance shifts as a representative of the polycentric paradigm, as well as it is clear that the term "territorial" brings not only more complexity to governance, but also a wider knowledge on the notion. However, understanding territorial governance is far from complete, especially in regard to the operationalization of the conception. In these circumstances, it is necessary to unravel the concept by following two steps. First, territorial governance could be explained in terms of its dimensions. For this purpose, reference is made to a number of authors that have dedicated their research to the definition and examination of territorial governance dimensions. Second, it is necessary to analyse these dimensions by verifying their existence and evaluating their level of effectuation within domains or sectors, where territorial governance is being implemented. Most of the research on territorial governance cases is made in the area of spatial planning (at various levels), with fewer cases developed around transportation and regional development (Schmitt and Well 2016a). The findings so far are promising, but their robustness would depend also on the possibility to revisit territorial governance in other policy domains.

In order to unravel territorial governance as a concept, Schmitt et al. (2013), Stead (2013), Böhme et al. (2015), and Davoudi and Cowie (2016) conducted analysis and empirical research to define the dimensions of [good] territorial governance. Referring to Davoudi and Cowie (2016), who develop their analysis based on the work of Schmitt et al. (2013), territorial governance dimensions are about input processes and not about the outcome of the territorial governance. Therefore, the dimensions make the concept more tangible by providing features that characterise territorial governance.

Schmitt et al. (2013) proposed five territorial governance dimensions in the ESPON TANGO project, based on the study of [good] governance concept and principles. For each dimension, they proposed 2–3 indicators (20 in total), which they initially validated for relevance and practicality through using a Delphi survey method, and engaging various experts. Further on they collected empirical evidence on 12 cases across Europe. Davoudi and Cowie (2016) revisit these dimensions in their chapter on "Good" Territorial Governance, which are then further analysed for another 20 European cases. The five dimensions are the following: (1) coordination of actions between actors and institutions; (2) integration of policy sectors; (3) mobilisation of stakeholders' participation; (4) adaptive to changing contexts; (5) realisation of place-based/territorial specificities and impacts (Well and Schmitt 2016, p. 13); (Schmitt et al. 2013, p. 12); (Davoudi and Cowie 2016). The first dimension is assessed through three indicators, namely: the governing capacity (organise, deliver, monitor, integrate), leadership (vision, ability to drive change, power), and subsidiarity as it is understood in governance studies in terms of linking decisions with the appropriate territorial/community scale of delivery (ibid.).

The second dimension is measured through *policy packaging* and *cross-sector synergy*. The first means bundling of policy areas/actions to add value and bring mutual reinforcement to governance processes, while the second implies synergies between actors in the different sectors (private, public, and civic) in favour of achieving sustainable development over a territory (ibid.). The third dimension is measured through *democratic legitimacy* (both representative and participative democracy), *public accountability* (of those who have the authority to make decisions), and *transparency* of the governance process (processes are open and visible to the public) (ibid.). The fourth dimension is operationalized through *reflexivity* and *adaptability*. Reflexivity "is about social learning and the ability to reflect on, review and revise territorially specific ideas" and actions, while "adaptability is about flexibility and resilience in the face of territorial changes and crisis" (Davoudi and Cowie 2016, p. 53). This dimension is about knowledge making, absorbing, and transfer for further use in the territorial governance process. This knowledge is helpful also to the existence of the final dimension with two indicators, namely: territorial *relationality* and territorial *knowledgeability* (Davoudi and Cowie 2016). In the first one, actors should be able to link the geographies of the problem with the solutions, without being confined by jurisdictional boundaries. The second implies the ability of using multiples sources of knowledge, and combining scientific with traditional knowledge for better solutions and better understanding of the policies' territorial impacts (ibid.).

Stead (2013) provides a slightly different list of, again, five dimensions underlying "the notion of territorial governance in policy discourse" (p. 142), namely: (1) coordination of policies/actors; (2) actors participation and consensus building; (3) powers and/or resources devolution to lower decision-making levels; (4) territorial cohesion; (5) assessment of territorial impacts and development of territorial visions (p. 142). Böhme et al. (2015) define also five dimensions, though with a less explicit formulation compared to the first two sources: (1) place-based approach and adaptive to changing contexts; (2) active participation of stakeholders, transparent, and consensus-oriented processes; (3) strategic and with long-term approaches to achieve societal objectives; (4) shared management approach and partnership principles for cohesion; (5) flexible and sensitive to territorial objectives and policies (p. 15/18).

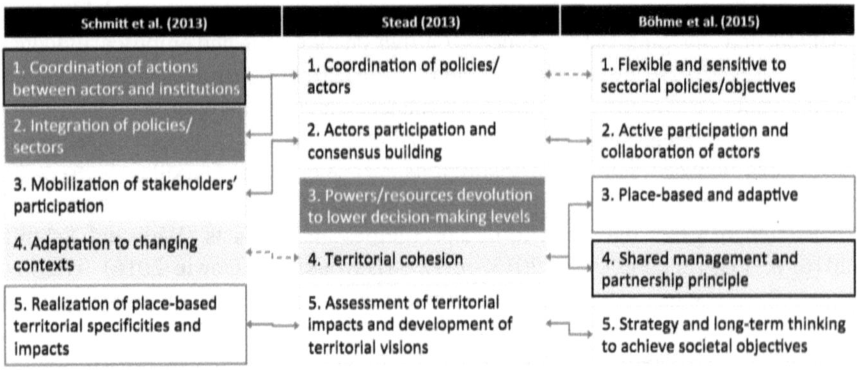

Fig. 7.1 Dimensions of territorial governance (source: author, based on Schmitt et al. (2013), Stead (2013), Böhme et al. (2015))

The differences between the three studies are in fact minor—mainly of an accentuation nature, and several overlaps could be identified, which in fact increase the robustness of the proposed frameworks for analysing territorial governance. Certainly, these frameworks help to understand and unravel territorial governance, but do not intend to show the quality of territorial governance, or its ability to deliver sustainable territorial development. Achieving the latter, it is necessary to undertake empirical observations of the successful cases and codify commonalities. Nevertheless, the presence of the dimensions is a good indication of whether the transition from multi-level governance to territorial governance is happening at all.

The first and the second dimension in Schmitt et al. (2013) are fully matched by dimension 1 in Stead (2013). Böhme et al. (2015) cover as well sectorial policies integration through the first dimension (Fig. 7.1), but add to the dimension also sectorial policy sensitivity in territorial matters, or vice versa, emphasis the need for coordination between sectors (vertically) and their territorial impacts (horizontally). Dimension 3 in Stead (2013) may be considered as a subset of dimensions 1 and 2 in Schmitt et al. (2013), especially if considering that the latter have subsidiarity as one of the indicators for measuring dimension 1. Thus, decentralisation and power devolution are a form of stakeholders cooperation, if we refer to governance shifts as explained, for instance, by Arnouts et al. (2012) or Araral and Hartley (2013), where power allocation happens in a top-down fashion from the government to lower levels. Still, this process of power transfer employs a crucial principle of multi-level governance—subsidiarity, according to which power is allocated to the lowest level (in territory and governance) where the function can be efficiently delivered. Hence, this principle has the ability to reflect the connection between actors' tiers and territorial contexts, which increases the relevance of appearing as a separate dimension in the Stead (2013) framework.

Stakeholder participation/mobilisation/collaboration and consensus building appear unequivocally as a separate dimension in all three cases, with almost the same formulation. An implied link exists also between dimensions 4 in all sources.

Hence, shared management and partnership principle are factors in achieving territorial cohesion, while the resilience of territories and institutions is a measure of territorial convergence and of shrinking spatial disparities. According to Davoudi and Cowie (2016), the place-based character and adaptability are the two key dimensions in Schmitt et al. (2013) that confirm the leap from multi-level to territorial governance. Likewise, though Stead (2013) does not use the same formulation, he calls attention to territorial [impact] assessments, development of territorial visions, and achievement of territorial cohesion as those making a distinction between regular governance and territorial governance (Stead 2013) and being explicitly a reflection of the place-based approach of governance. Finally, Böhme et al. (2015) point out the "place-based" as the key dimension for policy-making in territorial governance. According to them, such a policy can be adaptive to changing contexts by taking always into account place specificities and having the ability to be forthcoming in achieving long-term overarching societal objectives.

Territorial Governance of Common Pool Resources

Making a case on the importance of the territorial scope in governance shifts and on the territorial governance dimensions as a means to operationalize the concept steers logically towards the need for assessing the dimensions within policy domains and governance processes. The argument of this chapter is about the governance of common pool resources (CPR) as one model for territorial governance. There are three factors that evoke this hypothesis: (1) Territorial governance is the subject matter that exclusively addresses interactions between actors/institutions and the territorial capital. This complex set of interactions is differently known in environmental governance studies as human–environment interrelations, or socio-ecological interactions dealt with from a socio-ecological systems (SES) perspective. In the CPR theory, commons represent a thoroughly studied SES, which is capable of producing sustainable development; (2) Territorial governance deals with space and scale. Boundaries—fixed or fuzzy, jurisdictional or functional, contiguous or overlapping, as well as the size of the territory (scales) constitute the critical challenge of territorial governance—not simply for the various meanings, but mostly for the difficulty to implement the notion in sectors (for instance, in planning) and for the threat that boundaries' discussion may pose to democratic legitimacy—how to combine territorial representation with territorial governance (Faludi 2016). In the CPR discourse, boundaries and scales constitute as well as critical aspect of governance. On the one hand, there is a need for clear boundaries (physical and property rights) to consider a commons' regime as robust. Clear boundaries also add reliability to the assessment and quantification of actions and policies and to their legal consequences (Braat and de Groot 2012). On the other hand, commons as natural resources render ecosystem services to humans. These services extend their boundaries of effect much beyond the physical boundaries of the resource per se, therefore opening a discourse on rescaling the territory of rights and the measures to govern property relationships beyond the physical boundaries; (3) The commons' governance regimes exist since centuries and have proven to be successful in terms of sustainable outcomes. Reasons behind include robustness and adaptation capability of the institutions (therefore their resilience) as well as the place-based mechanism that these institutions employ. Therefore, the commons' regimes have the key dimensions that transform regular governance into territorial.

The CPR theory has a myriad of cases to refer to and draw conclusions from. However, this chapter will make use of the forest commons in Albania as an illustration. Albania is an interesting case for studying commons, due to its history of socio-political turmoil and abrupt changes of governance regimes, which have affected also natural resources and community institutions. Forests are chosen as the natural resource to illustrate the territorial governance of the commons for the following reasons: (1) forests have the highest supply of ecosystem services per land-use type and remain with the highest positive budget even after deducting demand for their services from each land use (Burkhard and Muller 2015; Luque et al. 2017; Moran 2005). This makes forests "givers" rather than "takers". (2) Forests cover around 30% of the planets land area (MA 2005b; FAO 2016) and provide a significant number of critical ecosystem services to humanity (United Nations 2014), while being the primary source local communities around the world use to sustain their livelihood. In this respect, the need for forests' ecosystem services is tremendously high at both, the local community and the global scales. Worldwide, forests face a severe deforestation and degradation trend with about 13 million hectares per year (FAO 2015b), therefore being responsible for about 15% of all greenhouse gas emissions (WWF 2017), due to reduced overall carbon sequestration capacity. Forests' role in climate change (a global threat) is therefore unquestionably vital, as it is their role in communities' livelihood sustenance.

7.3 Forest Commons as a Model of Territorial Governance

Key Factors in the Territorial Governance of Common Pool Natural Resources

Territorial governance is evolving as a response to common societal challenges felt on the larger scale, such as climate change, cohesion, and sustainable development (Stead and Pálné Kovács 2016). However, territorial governance, or as Young (2013) calls it, governance without government, happens also at the small societal scale, where natural resources are managed throughout the world successfully in common by local communities, i.e. their "direct" appropriators (Ostrom 1990, 1999, 2009a, b; Young 2013; Raymond 2003; Agrawal 2003). After all, "the geographic frame of territorial governance may be wider or smaller than the state unit, depending on the issue at hand" (Faludi 2016, p. 36). Most of the talk is about rescaling between local to regional, national, and global levels, but the presence of commons' regimes suggests for an even lower level of rescaling(Young 2013)—the one where socio-ecological interactions involving communities, are borne. This gives more sense to switching between "place-based" and "territorial" concepts. As Faludi (2016) says, "place is an umbrella term for spatial formations" (p. 37), regardless of the jurisdictions (ibid.). The governance of common pool resources is not a political pre-designation of powerful actors within politically set boundaries. It is a genuine peoples' organisation thriving on specific contexts, and fuelled by the complexity of socio-ecological interactions, perceived in relatively small territorial scales. It is argued to be polycentric governance happening at the small scale of common pool natural resources, and if a factor of nested enterprises is present than the scale of the polycentric interaction increases.

Institutions: Design Principles

Research on common pool resources' institutions has shown that the prospering cases carry in common a number of features, named by Elinor Ostrom as design principles. Ostrom identified eight design principles of robust commons' institutions: (1) well-defined boundaries (physical and property rights); (2) proportional equivalence between benefits and costs; (3) collective choice arrangements; (4) monitoring; (5) graduated sanctions; (6) conflict resolution mechanisms; (7) minimal recognition of rights to organised, and (8) nested enterprises. Some of these principles are related to the scale. For instance, the first principle is about the clarity of property rights that the users have towards the resource (Ostrom 1990, 2009b), as well as about the resource's boundaries (Cox et al. 2010). If the knowledge of the CPR system is to be transferred to the larger scale commons, then the complexity of this principle increases further to include time boundaries and the impact bearers (Toto 2018). In order to bring clarity to the first principle at least for the small scale of the commons, Ostrom proposed a bundle of rights metaphor, where access is the basic level of rights that a resource user has, while alienation—the right to sell and/or transfer the property to others, is the highest level and completes the full set of rights added in a cumulative fashion (Fig. 7.2). The user that does not have alienation rights is called a proprietor.

The last design principle-nested enterprises has also a territorial dimension. According to Ostrom, though CPR systems are often small, there are also cases where several commons' institutions interact among them and in complex socio-ecological contexts, thus governance actions take place within multiple nested layers. This design principle is not always present, but when it is, it leads to the development of a polycentric system of governance for larger scale commons. In the case of forest commons in Albania, this principle is present through 256 forest users' associations organised on micro-watershed level. Each association represents the various groups of commoners within the micro-watershed on a policy discourse level and strives to ensure access to funding for supporting the commoners with forest maintenance activities. All associations cooperate among them on a watershed basis and with the national forest federation mainly for influencing government's policy on forests.

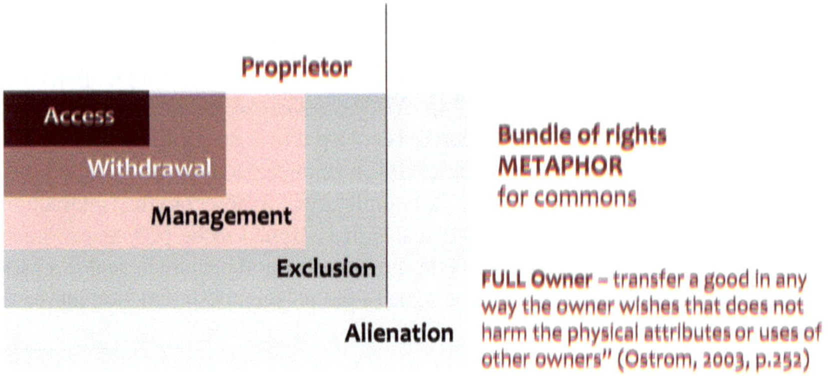

Fig. 7.2 The bundle of rights metaphor for common pool resources (source: author's interpretation based on Ostrom (2003))

Scale

Another finding of the international research on commons is that scale is an important factor for robust commons' institutions, which have usually a very small commonly owned/managed space (Kluvánková and Gežík 2016; Ostrom 2009b; Van Oel et al. 2009) compared to the jurisdictional space. The previous is defined by the edges of the related natural resource and the rights of the owners/users on the resource, while the latter is a political designation, where ecological criteria are hardly factored in. However, the scales discourse is not a closed issue, as there are studies that argue both, in favour of and against the small scale (Agrawal 2000), or size as a factor for commons' governance success. There is though agreement that commons' systems exist since centuries (Ostrom 1990, 1999), are sustainable, and succeed based on tradition and knowledge passed down to generations and within families. Nowadays, while governments are dealing with global environmental threats, the society takes it for granted that the management of the commons at the small scale is ceaseless and bears no risk of impact from the global problems (Kluvánková and Gežík 2016; Merino 2015). The concern on the reality of global commons (Merino 2015) and the level of uncertainty that sustainable development has to deal with (Handmer and Dovers 1996; Palahi et al. 2008) are increasing and showing that the links for transferring knowledge from the successful small scale to the failures of natural resources management at large scale are probably weak, or not present (Young 2013). This leads to a failure in bridging between scales, when it comes to the governance of natural resources with a territorial perspective.

Stead and Pálné Kovács (2016) highlight that "actors gain or lose influence as a result of authority being reconfigured around new spaces and territories" (p. 23). Keating (2009) calls this as processes of rescaling, which are by no means uniform across the territories (Stead and Pálné Kovács 2016). However, the territorial influence is not lost with time, it merely shifts either from one territory to the other, or through generations, to appear again. For instance, because of severely abrupt changes overtime, due to exogenous reasons—dramatic ideological shifts, and endogenous reasons, i.e. stakeholders' lack of experience and coherence (Pejovich 1990), the forest commons in Albania "ceased to exist" for almost half a century. Historical ties within families and traditions flowing through generations, made it possible for this governance system to return back and thrive after 1990, even though informally. Apparently culture and traditional knowledge must have played a crucial role in transiting influence through generations and across territories. Besides, as people safeguard natural resources and territories based on a specific values' systems of reference (Raymond 2003; Kareiva et al. 2011; Chan et al. 2011; Goulder and Kennedy 2011; Brnkaláková 2016), the community's influence over these forest territories was regained also through a values' transfer process. This raises the necessity of examining the sets of values transferred in different contexts and territories, in order to understand how territorial influence is maintained/transferred in various settings, and whether this can teach any lesson in regard to the above scales' failures.

Values

In order to achieve its pivotal role, that of achieving sustainable development (Meadowcroft et al. 2012; Lange 2017), governance has to "reconcile economic effectiveness with social cohesion and ecological equilibrium" (OECD 2001, p. 25). Hence, sustainable development is a value-oriented debate and struggle, focusing on how to promote equity and justice within and among generations from a socio-economic and environmental perspective (Derissen et al. 2009; United Nations 1987). The United Nations Bruntland report (1987), where the concept was originally formulated, defines two sets of values that shape human demand for resources and prosperity: on one side there is utility, overall welfare, and poverty reduction as key drivers, and on the other there are equity, moral, ethical, cultural, aesthetic, and purely scientific reasons for conserving wild beings. People should model their existence by establishing harmony between both sets. For this to happen, political systems and institutions should develop in such a way as to promote inclusion and democracy, avoid fragmentation, create healthy socio-ecological relations, endorse actors' cooperation at any level in the society, and adopt governance modes that are able to cope with uncertainty (United Nations 1987). This leads to the establishment of socio-ecological systems which are capable of withstanding or adapting to changes indefinitely, therefore being resilient (Handmer and Dovers 1996; Derissen et al. 2009; Perrings 2006). Resilience, measured through adaptability, self-learning, and robustness to change (Perrings 2006; Armitage 2008; Nagendra and Ostrom 2012; Schiappacasse and Müller 2018), is the way to think about sustainability (Perrings 2006) and therefore a fundamental conception in governance. Going back to the territorial governance discussion, these features of resilience constitute also the key dimensions that give governance its territorial perspective as defined by Schmitt et al. (2013), Stead (2013), and Böhme et al. (2015) and qualify it for the shift towards polycentrism. However, digging further into the resilience discourse is not an aim of this chapter, as it is instead discussing the values that stand behind any policy decision-making, which affects the resilience of the common natural capital and its governance.

The discussion on values is primarily a discussion of anthropocentrism versus biocentrism. Utilitarianism is currently the worlds' dominating doctrine with a focus on the aggregate satisfaction of the human society and no consideration of egalitarian principles (Alexander and Penalver 2012) among humans, let alone between humans and the natural capital. The latter is also reflected in the failure of governance to address the values of natural capital (Hysing 2009) form a holistic perspective, i.e. from instrumental ones—serving to the humanity, to inherent values—those of the ecosystem for itself (Beatley 1994; Kareiva et al. 2011). Therefore, an evaluation process that considers the full range of values is still far from being accomplished. However, the examination of the links between ecological and socio-economic systems has advanced significantly as to generate knowledge that feeds policy advice and decision-making, raise global awareness, and teach nations the importance of undertaking integrated ecosystem assessments as a standard step in governance (MA 2005a). The largest and most significant study on the status and trends of the ecosystems in the world—the Millennium Ecosystem Assessment (MA) has articulated also the major gaps for further research with regard ecosystems service values. These gaps include, among others, the limited capability of environmental and ecological models to incorporate spatial, ecological, and institutional feedback in real-time (Shoeibi et al. 2015).

The European Environment Agency classifies ecosystem services—ecosystem functions that turn into services at human demand (Kareiva et al. 2011; Haines-Young and Potschin 2011) in three groups: (1) provisioning (for nutrition, materials, and energy); (2) regulation and maintenance (mediation of waste and flows and maintenance of ecosystem conditions); and (3) cultural (physical, intellectual, spiritual, and symbolic interactions) (Haines-Young and Potschin 2012). In this myriad of services, humans tend to value, and therefore consider in governance, those that are easily apparent and provide tangible benefits—mostly the provisioning services and some of the cultural-aesthetic values. Even in this case, there are services, such as renewable energy, or recreational activities in protected areas, that may not receive proper valuation, as the latter happens through market methods. In these methods, due to lack of knowledge on the full range of services that the resource can provide, the commodity price does not necessarily represent the full value of the resource. Of course, valuation of the regulating and supporting services (either indirect to humans, or not easily perceived by them) and cognitive development services (such as pride on the resource and symbolic interactions) is even more complex, not so much due to lack of valuation methods, rather than because the resource users lack knowledge on the existence of these services and/or are not aware of achieving cognitive development from the natural capital. The values that resource users assign to their common natural resources affect their way of natural commons governance, and the resource-related know-how is clearly a factor that impacts the values' system.

Besides knowledge on the ecosystem services, values depend also on the knowledge on scales dependencies, which is incomplete (Grunewald et al. 2015). Ecosystem services have a physical scale of impact that exceeds the geographical boundaries of the common pool resource and hence also the boundaries of the appropriators, resulting in a scale mismatch between human rules and ecological realities (Low et al. 1999). A typical illustration to this is that of forests and water within a river basin. Usually, those who live downstream benefit from the water regulation services that forests provide, or may be negatively impacted if upstream users do not maintain the irrigation system. Those who live in urban, agriculture, and non-forest areas benefit from natural pollination from species that breed in the forest, and air and climate regulation, as well as erosion control provided by forests. Thus, while a small group of commoners govern the forest in common, a larger and undefined group benefits or bears impacts merely because natural resources and their services exist. This larger group has no legally regulated rights on that specific forest, has no knowledge on the services it is benefiting from, is not aware of the scale mismatches between ecosystem services provision and decision-making, has no role or contribution in the forest's governance (Satake et al. 2008), and therefore assigns no value whatsoever to the natural resource. This is a critical and complex rescaling issue that forest/natural resources governance has to deal with in order to guarantee the sustainable development of the resource at both, the small (community-local) and the large (regional-to-global) scale.

Obstacles

The rescaling discussion in natural resources and ecosystem services governance is addressed so far mainly through research on biophysical structures/processes characterisation, ecosystem services valuation tools, and through attempts for mapping ecosystem services provisioning and benefiting areas (Maes et al. 2012). The research has made significant advances in regard to mapping supply of ecosystem services, while the mapping of benefiting areas rests mainly with land-use mapping and attempts to calculate their demand for specific services at global scales (for instance, Corine[4] land uses land cover is mostly used instead of detailed local land uses and their specificities). The drivers and the pressures, i.e. the values' systems that generate the demand for ecosystem services, are not studied and mapped—probably because this is a very complex exercise. In these circumstances, the territorial governance of natural common pool resources is faced with three possible obstacles:

[4]According to the European Environment Agency, the Corine program was initiated in the European Union in 1985. It offers among others a database of the inventory of land cover in 44 classes, and presented as a cartographic product, at a scale of 1:100,000. This database is operationally available for most areas of Europe and is often use in EEA and other institutions' studies and research.

1. The rescaling between governance levels and modes is underexplored. Hence, generalising or transferring knowledge from the small sustainable scale of common pool resources to the larger scale of commons is yet lagging behind (Young 2013). It was already argued that rescaling is not an issue to ignore because it reinforces the territorial approach to adaptation and resilience, which is necessary in a context where global threats and related responses are linked directly to their local manifestations (Bertrand et al. 2016, p. 195). On the other hand, local responses or governance models embody the multitier and place-based character that ensures resilience, which global policy initiatives were less able to develop and address (Rametsteiner 2009, p. 143);
2. Knowledge transfer from experts to communities and vice versa is also a rather quiescent process. The mutual exchange between scientific and traditional knowledge, for the sake of addressing ecosystem services in governance, would generate knowledge dynamics for localities to fully benefit in an increasingly global era (Halkier 2012), while being also a solution to the rescaling problem. The commons' scale performs based on traditional knowledge, while policy decisions at the large scale are taken based on scientific knowledge.
3. Ecosystem services value and human demand for them are poorly addressed in governance (Hysing 2009); (Grünwald and Wende 2015). Research wise, most of the focus is placed on generating know-how about ecosystems as biophysical structures and service providers. This process is not well balanced by know-how on the demand that the society poses to ecosystems, the fundamental reasons that stand behind, the values that feed demand, and the way for identifying, measuring, and comparing demand with the supply.

Methodological Aspects

In order to study the territorial governance of the commons, the following case of forest commons in Albania takes a two-phases approach. First, a model is built to combine in one single platform all the components of commons' governance as well as the three obstacles that sustainable territorial governance of natural resources may face across territorial and governance scales. This model brings together governance modes with territorial/ecosystem knowledge and human demand (values) in a closed circuit. The model represents both, socio-ecological interactions, as well as a platform for analysing territorial governance of the natural resources. Second, an assessment of the territorial governance dimensions for forest commons in Albania takes place. In order to assess each of the 12 indicators representing the five dimensions of Schmitt et al. (2013) as well as cover the sensitivities that Stead (2013) and Böhme et al. (2015) add with their dimensions, use is made of the model build in the first step (Fig. 7.3). Finally, from a spatial perspective the case if focused in the commons of one river basin. The river basin is a naturally occurring territory—a larger scale common, with an entirety of ecosystems and socio-ecological functional links, where the scales' mismatch for commons and their services can be

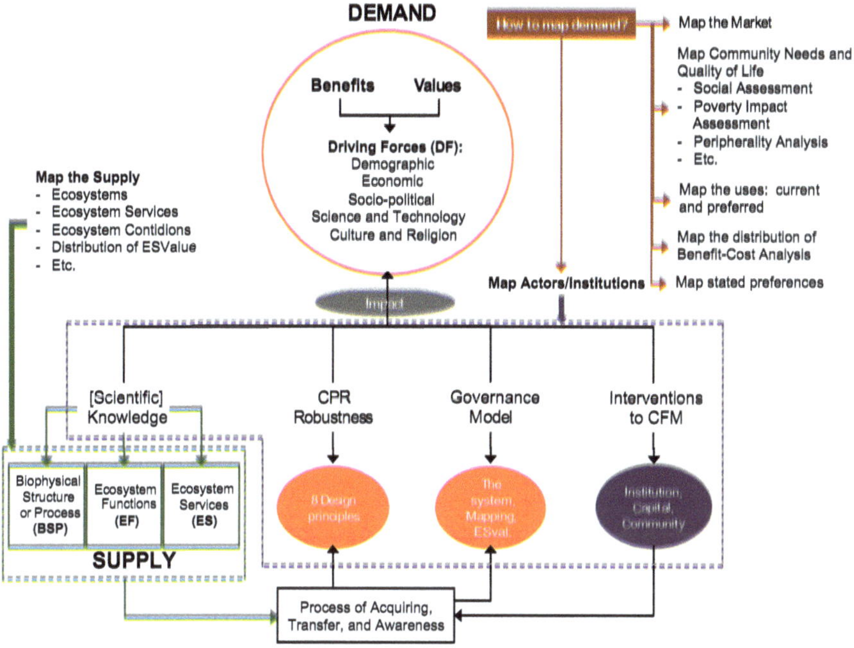

Fig. 7.3 A platform for achieving territorial governance of the natural resources (Toto 2018)

studied without implications from ecosystem services offered in the adjacent watersheds.

Methodologically speaking, the following summarises the steps for assessing the territorial governance of forest commons in Albania, based on the above model:

1. The governance model in terms of institutions and interactions is assessed on two levels: The commons and the conventional forest governance. Forty-four variables are analysed, out of which eight are the design principles of robust CPR institutions developed by Elinor Ostrom. For the purposes of this case, reference is made to the unravelling of the design principles proposed by Kluvánková and Gežík (2016), applied also on Slovakian forest commons. All the remaining variables are taken from Fleischman et al. (2014/a) and the SESMAD[5] database https://sesmad.dartmouth.edu/variables. This project intended to explore if variables found important in explaining outcomes on small-scale systems could be scaled up to explain outcomes in large-scale multitier environmental governance (Cox 2014). Various studies show that key variables that may have influenced a process happening to a CPR (such as deforestation rates, or other endogenous and exogenous context-based variables (Armitage 2008)) are not well captured in CPR theory. These variables include, among others, the intention of the government's system often recognised as the "government's practice or intervention"

[5] Socio-Ecological Systems Meta-Analysis Database

(Foucault (1991) in (Li 2006/2)) that attempts "to shape human conduct by calculated means" (Li 2006/2, p. 3); the presence of clientelistic politics; the influences of international politics and markets (interventions as described above); the influence of top-down governance (Fleischman et al. 2014/a); the socio-cultural construct of the stakeholders group, and their ability/intention to self-organise; and the specific knowledge on the ecosystem and on the ways the ecosystem can shape the human behaviour (Armitage 2008). To cover this variety of factors, the SESMAD project has developed a large set of variables and indicators, applied to a number of cases worldwide. Still this is pioneering research and needs to enlarge further the repository of cases for producing generalised findings. For the Albanian large-scale forest commons a selection of 36 variables is made, choosing those that best fit with forest management processes. Collection of information has taken places through 11 focus groups organised within one river basin and 20 interviews with forest officials at municipal and national government levels.

2. A mapping of ecosystem services supply and demand budgets for Corine land uses is made based on the methodology of (Burkhard and Muller 2015). Thirty-one maps showing the distribution of the ES supply-demand per each ecosystem service are prepared. A final spatial interpretation of human pressures and environmental hazards in the basin versus forests is made (Figs. 7.4 and 7.5).
3. The overlay of the information from both first steps is interpreted through the analysis of driving forces, pressures, state of the environment, impact, and responses (DPSIR) for forests in the basin.
4. The elaboration of the five territorial governance dimensions is made on the basis of the information collected through the first three steps.

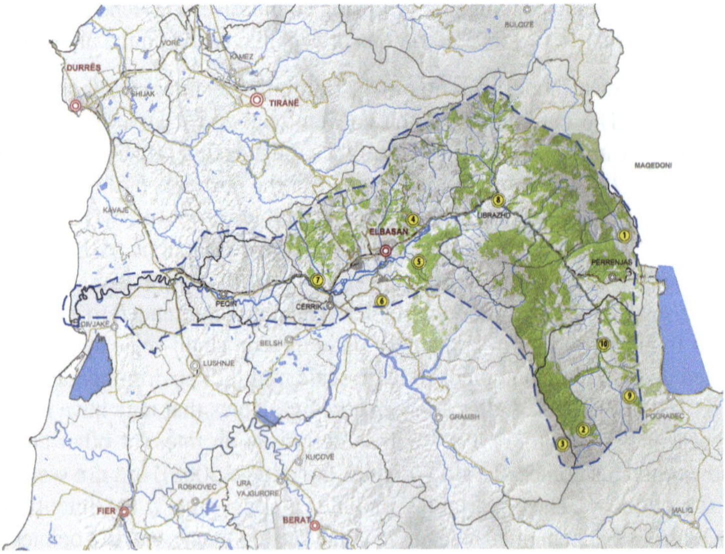

Fig. 7.4 Forest commons in the Shkumbini river basin (source: author based on Corine 2012, focus groups and visual survey)

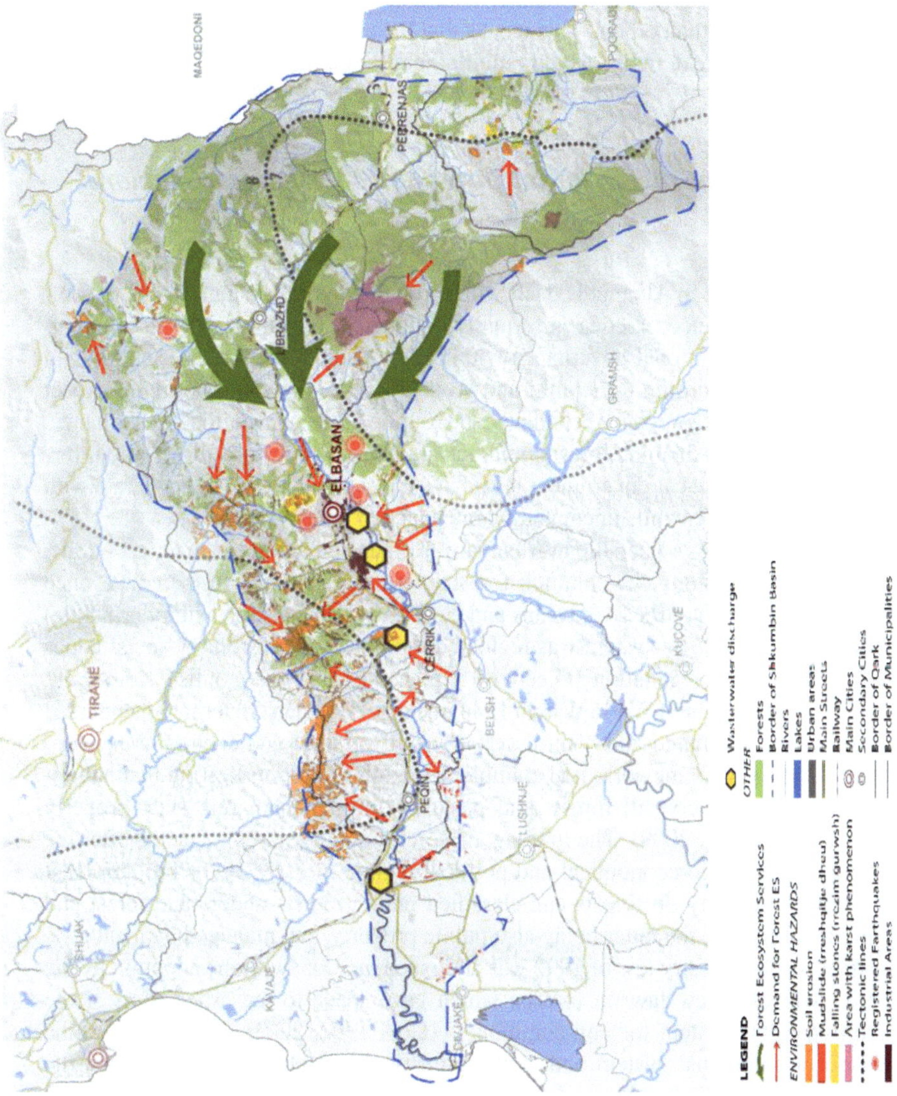

Fig. 7.5 Forests versus sources of pressures in the basin (source: author based on Corine 2012, SHGJSH 2015, and own graphical interpretation of the ES supply-demand budgets)

The model and the methodological steps described above can be used in the study of any context of forest commons, which is aiming at exploring territorial governance. However, as Davoudi et al. (2008) argue, considering that territorial governance has different dimensions and there are context and institutional framework specificities in different policy domains and territorial scales, it is necessary to recognise these specificities and understand that for each level, some dimensions are more relevant than others. This means that the application of the model to other commons or policy domains would require the relevant modifications.

The Dimensions of Territorial Governance for Forest Commons' in Albania

Forest governance in Albania is dealt with through a hybrid system, where the soft territorial governance practice, informally conducted, is combined with interventions of the conventional governance within the jurisdictional spaces (Schmitt and Well 2016b). According to Schmitt and Well (2016b), one would normally expect that the functional approaches challenge the routine of the conventional governance. However, though Albanian forest commons are not based on forest legislation, forest officials at the local government level have chosen voluntarily to cooperate with forest users in the establishment and maintenance of forest commons.

Till 1945, forests ownership in Albania was organised in three categories: public, private, and common. The common forests were those located in the conveyance zone (Marsh 2010) of the river basins and close to the villages (proximity principle) (Raymond 2003). The system was regulated through both, customary social codes and regular state legislation (Gjeçovi 1925; Ministria e Ekonomisë Kombëtare 1930). Soon after the II World War, in 1944, the new Albanian Government endorsed a centralised command-and-control system of governance and economic development. Under this framework, it also implemented the Agrarian Reform that nationalised (among others) all forests and pastures, turning them into state property (Muharremaj et al. 2009). The first law on forests [and pastures], enacted after the change of the socio-economical and political regime of early 1990s brought back the private property on forests and classified public forests under state forest and communal forests, the latter being also public property, and managed by local governments. The forestry law of 1992 was abolished in 2005, with the adoption by the parliament of a new law on forests, which gave space to the articulation of the licenced rights system for non-commoners (GoA 1992, 2005). In response to the lack of a commons' system and to the priority given to the licenced rights, the National Forests Federation and the forest users' associations established in mid-1990s initiated in 2014 a lobbying and advocacy process to influence the third wave of forest legislation, which is still going on.

In 2015, the Government of Albania enacted a forest moratorium law aiming at halting the exploitation of timber forest products for internal commerce and export. Timber exploitation for heating purposes (managed by the municipality for the needs of its own residents), conversion of forest land-use into other land-use categories, and forest cutting for maintenance purposes are however allowed activities. The moratorium in itself is quite controversial, because of lacking further bylaws to regulate its main provisions, being enacted without prior assessment on possible external effects, and with no provision of other energy alternatives. Furthermore, the moratorium enforces a "simplified" one-rate penalty system of around 45,000 EUR per violation, regardless of the violation type and size (GoA 2005). This is in contrast to the previous graduated sanctions system and has in fact led to either unjust sentences, or to an increase of transaction costs due to higher corruption costs.

Though the government has not recognised forest commons by law, it has allowed room for two significant donor' interventions (World Bank and USAID) to re-introduce and support a forest commons' system, as well as for other initiatives that support forest users' associations and their forest maintenance activities. The two first projects aimed at sustainable community forest management, poverty reduction, and rural livelihood improvement. However, one chose to work with communities, and established the network of forest users' associations (256 to date). The other worked with private forest owners. Both were successful during the project implementation lifetime because of the substantial donor technical and financial support, and their leverage power with public institutions, as well as engagement of several experts and NGOs. However, the momentum gained in regard to the establishment of a forest commons system was lost once both interventions ended and a system of commons could not be introduced in the legislation, though it was informally re-established. In 2015, the government of Albania revised its decentralisation reform and as a result forest governance was devolved to the local governments. Currently, municipalities own 82% of the Albanian forests, the national government owns 15%—the national protected areas (National Agency of Protected Areas 2015)[6] and the private owner have 3% of the forest coverage (Muharremaj et al. 2009; Muharremaj 2003; INSTAT 2016). In reality, around 30%[7] of the local forests are managed through a commons' regime. The forests currently managed in common are located close to the villages and are those owned in common prior to 1945. Hence, the allocation of rights after 1990 has happened based on traditional and historical ties with the forests and knowledge transferred through generations within the families. Each family is assigned a part of the forest, proportionally to the size owned by the families before 1990. Each family manages individually its own portion of the forest, but monitoring is an activity they all carry out in common.

Coordinating the actions of relevant actors and institutions: Besides being supported by donor projects during the first 2 decades after 1990, the forest commons'

[6] The area is calculated in GIS based on the specific data provided by the National Agency of Protected Areas for each specific area.

[7] This is not an official figure. It is based on the fieldwork (interviews, meetings, focus groups) undertaken by the author for purposes of research on forest governance. The calculation is conservative and the figure is suspected to be higher.

thrive in Albania because of two other important factors: (1) the governing capacity of the government's institutions is insufficient to deal with the complexity of forest ecosystems, due to lack of human and financial resources and low recognition of ecosystem services in forest management. Under these circumstances, municipal officials have decided to cooperate with the commoners in a win-win solution, trusting their traditional knowledge, and genuine interest on protecting and maintaining forests; (2) the commoners have minor benefits from the provisioning forest ecosystem services because by law they cannot withdraw and/or sell forest products. Still, they safeguard and maintain the forests based on family pride, thus based on cultural and cognitive development values.

The latter have survived through generations and constitute the key factor in the survival of forest commons governance in Albania. This shared value ensures a bonding between commoners and the resource, though there is no leadership to guide their actions, while they manage the respective portions of the CPR individually. There is though guide and support from the local officials and the forest users' associations. The latter gather the commoners to discuss issues of common interest and share the information in layers of cooperation with the other associations and the national forest federation. The system is decentralised and the subsidiarity principle is implemented in the devolution from the national to the local level of government. By "informally" transferring functions to the commoners, the municipality is de facto implementing the subsidiarity principle through "allocating power" to the most appropriate level of decision-making—the community. Furthermore, the users' associations take also responsibility on forests' governance. They guarantee the network governance that transfers knowledge among layers, while representing the commoners' rights in higher levels of policy-making. However, there is one risk on the performance of the system and that rests with the legal basis for the commons: neither common property nor commons' regime are recognised by the legislation. This increases the uncertainty on forest commons, therefore putting at jeopardy the long-term stability of this "informal" solution.

Integrating relevant policy sectors: This is probably the weakest dimension in the territorial governance of the Albanian forest commons. There is some level of cross-sector synergy or as (Davoudi and Cowie 2016) call it: horizontal coordination and cooperation between public (municipalities and the ministry), private (commoners), and civil society (users' associations and the national forest federation) to work in favour of the sustainable forest commons governance. However, only the conventional governance system makes forest management and use decisions. Formal governance includes the Ministry responsible for forests, the state agencies responsible for forest protected areas, and municipalities. The instruments and regulations that these institutions refer to for governing forests are rooted in the legislation. The law defines that forest management should take place on the basis of the provisions made in local forest management plans. Each plan has a life expectancy of 10 years and should be revised after this period. Land conversion on the other hand takes place through the territorial development plan, and it is effectuated through the minister or government's decisions, depending on the area of the required forest conversion. The commoners have no say in this process, as they have no rights to interfere in case the government allows exploitation of water and energy resources within the forest. The forest sector, both in the municipality and in the ministry, does not cooperate with the other policy sectors when preparing policy documents or management instruments. At the large scale, conventional forest governance remains largely sectorial. Yet, spatial planners carry out analysis on the quality and spatial distribution of forests, and based on the demand for development, propose forestland conversion or protection. Depending on the governing capacity of those preparing the plans, the analysis may be comprehensive and scientific, or rudimentary and superficial. That will then also impact the quality of the decision-making.

Mobilising stakeholders' participation: This dimension is analysed separately for any of the two governance modes. In the case of the conventional governance, legitimacy is ensured through representative democracy. The mayors and municipal councils are locally elected and represent the interests of their constituencies in various domains, among others also for the forestry sector, which is a decentralised function. Let's reiterate that municipalities are responsible for and own 82% of forestland in Albania. The case of forest commons, on the other hand, is an example of participative democracy, where the commoners are free to make their own decisions as a group, within the spectrum of rights "allocated" to them by the municipal officials.

The public accountability discussion is also specific in each case. Thus, on the one hand there is the case of forest commons, where accountability is high as long as the local officials have decided to embrace the commons' system and cooperate with the forest users, showing thus care and concern for the forests. On the other hand, the municipalities, with insufficient governing capacity (as argued above), take limited care of non-common forests. These forests are located in the higher altitudes of the river basin, which according to Marsh (2010) constitute the contributing zone where most of the water catchment takes place, and most of the forests' services are generated. Decision-making does not necessarily reflect the context and monitoring and enforcement are weak for forestland in these higher altitudes. Finally, transparency is also a two-face coin. So, processes related to the commons are fully transparent, whether being undertaken by the commoners or the municipality. Processes on the non-common forests go usually unnoticed. The licenses are provided through procedures defined in legislation, but in absence of information for the local communities. Forest conversional so takes place through top-down decisions and it is enacted with little or no information whatsoever for the local communities.

Being adaptive to changing contexts: The context for forest governance has experienced dramatic changes overtime, with the shift from a mainly rural economy to a centralised command-and-control regime in 1945 and then long transition since 1990 to a decentralised and market-oriented socio-political system. These governance shifts were followed by changes in forest ownership and property rights, which influenced also peoples' attitudes towards forests. The major impact was and remains deforestation that is thought to be at the rate of more than 20% in the last 27 years.[8] It is logical to think that in a disturbed and unsteady context of governance and property regimes, the connection between forest users and the resource is to be lost. However, as it was already argued, not only a forest CPR regime is present in Albania; it is also performing sustainably at the small scale. As a matter of fact, the strongest ties with forests are created in those villages, where the historical memory of forest ownership (mostly proprietary and in few cases also alienation) is still vivid. The users show high interest in protecting the forests because 80 years ago, the forest belonged to his/her family. Hence, though users have a moderate livelihood dependency on the forests they take care of, and commons were supressed for 50 years, still collective action has thrived back based on three factors: cognitive development values are strong among commoners; the inherent link between users and forests (as property and as a natural resource) remained almost intact due to transfer of traditional knowledge through generations; and external interventions to the governance regime through donor projects have promoted the reintroduction of the commons system. In these circumstances, the group of commoners can be described as of high adaptive capacity. The practice of organising as a group around commons has somehow weakened, but commoners have found new ways to approach forest commons governance, by subdividing the common forest into family shares and carrying out together the monitoring of the forests. This is in fact a pattern of system's reflexivity, as the ability to learn from and reflect on the context changing dynamics. Reflexivity is also present in the voluntary engagement of the commoners in the reforestation of areas that were previously deforested for agriculture use and orchards, as well as in their eagerness to learn and adopt new sustainable forest management practices.

The adaptive capacity is high also due to the diversity of income, which is not based merely on forest user rights. As mentioned earlier, the commoners value much more the spiritual and cognitive development services they get from forests than the provisioning ones. According to the fieldwork, they mostly base their life sustenance on agriculture and pastoralism, trade and services, remittances, and to a lesser degree on administrative works and construction.

Realising place-based territorial specificities and impacts: The duality observed so far in the governance system description is observed also from a spatial extent perspective. The top-down approach, aka the conventional forest governance, is organised around protected forests areas, with a geographically set spatial extent varying from few hectares to some km squares and distributed over the country; and around forests owned by the local government, within their administrative boundaries, organised in 61 municipalities. Prior to the territorial and administrative reform of 2015, Albania had 373 local governments. This administrative territorial structure was very fragmented, but organised around sub-watersheds (in most of the cases) and with smaller size territories. In these conditions, managing forests was rela-

[8] Apart from Corine 2000, 2006, and 2012, there are no official records of the forest cadaster. The figure results from calculations made based on Corine data as well as from interviews and focus groups with forest officials and forest users.

tively easier and more ecologically fit. The significantly increased size of the new municipalities is expected to prove efficient in terms of public services, but not in terms of forest governance. While the size of the forest territory to be managed locally has increased by manifolds, the number of forest officials has reduced drastically. Furthermore, forest distribution and ecological interactions were not considered as a criterion of territorial subdivision/merge in the territorial and administrative reform. As a result, the larger size of the municipalities has not managed to internalise the boundaries mismatch issue in ecosystem-based forest governance.

The spatial extent of the forest commons governance has a different character from that in the conventional one. Forest commons are dispersed over the territory, have varying sizes from 1.5 ha to 200 or more ha, and follow the pattern of rural settlements location and fragmentation. This system is not dependent on administrate boundaries. Still the boundaries mismatch is evident in the case of forest commons, when the scale increases to include also the services they provide, which exceed the scale of the common. The illegal exploitation of forest commons by external users (out of the commoners group) mainly for heating purposes is an indicator of the risk that commons' boundaries may bear. The forest users' associations and the municipal officials play a role in the facilitation of interactions among commoner' groups, regardless of the jurisdictional boundaries their forests belong to.

Territorial knowledge ability is also good with respect to the traditional know-how that commoners have and use in forest management activities. They have real-time information on the situation of their forests through implementing a collective forest monitoring system. However, the users' scientific knowledge is low and mainly intuitive, based on family knowledge transfer. This may set a pressure on the resource, even though the users are willing to protect it for the long-term, because the ecosystem-based governance happens ad hoc and based on inherited know-how. The knowledge of the forest experts at local and national level, on the other hand, is above average, but cannot be considered as a typical pluralistic one. Municipal forest officials have mostly a "one-resource management" perspective and are less aware of the diversity of interrelations between ecosystem services and human needs.

7.4 Conclusions

The increasing complexity of the society and its radical changing leads to governance shifts which, besides indicating a status of social and institutional evolution, are challenged by the emergence of new knowledge dynamics (Kovač et al. 2016; Halkier 2012; Driessen et al. 2012; Arnouts et al. 2012; Rametsteiner 2009; Rhodes 1996). Polycentricity is the governance shift the society is heading towards and territorial governance is the way to address it, by being multi-scalar, multi-sectorial, and multi-actors. Polycentricity, with its unique design of handling local and global challenges across scales and in increasingly complex societies (Araral and Hartley 2013), has the ability to assure knowledge transfer among multiple actors. Its implied interaction of various levels of government with non-government institutions facilitates markets and lowers transaction costs, thereby achieving the purpose of governance. Polycentricity emphasises the importance of territory as a dimension in the governance conception. This is to say that the territorial dimension is not a new invention, but an inherent indicator of a logical shift that builds on previous governance modes, while aiming at widening their scope, so as to enable them addressing the increasing number of stakeholders, issues, and socio-ecological interactions.

This chapter argues that territorial governance happens in and through other policy domains, which interact among them and with the territory in several different ways, to produce territorial development. Territorial governance moves away from the sectorial development approach, which creates spatial inequalities, does not enable collective action, is not place-based, and has a rather limited level of adaptability. Most of the studies in territorial governance take a social-institutional or spatial planning perspective. This is probably because most authors are concerned with providing a proper and widely accepted definition of territorial governance, analysing its dimensions and the evolution of the concept; while others take impetus from the term "territorial", which is closely and visibly linked to spatial planning. Despite the fact that these accounts, though explored, are still far from being fully researched, the investigation of other understanding mediums in territorial governance is imperative. This helps in creating an all-encompassing view on the concept. Collective action and socio-ecological resilience as the direct outcome of territorial governance enable sustainable territorial development and are best seen through the study of socio-ecological interactions.

But, territorial governance needs operationalization, and this is best done through the analysis of its dimensions. Schmitt et al. (2013), Stead (2013), and Böhme et al. (2015) propose five dimensions, with the purpose of measuring territorial governance processes. The application of these dimensions is appropriate for large and for small-scale territorial governance. As territorial governance happens across scales assessing its dimensions provides good indications on how to transfer and generalise local knowledge for solving global societal challenges. The study of the forest commons as a model of territorial governance is helpful in two directions: (1) unravelling territorial governance at the small scale, thereby increasing the understanding on the territorial governance conception as place-based, bottom-up, adaptable, and with territorial insight; (2) dealing with the scales mismatch, or the rescaling challenge and drawing lessons that can feed the larger scale policies, through emphasising knowledge transfer/exchange and the need to employ a larger spectrum of values in governance.

The case study on forest commons in Albania validates that by employing a value system that goes beyond the utilitarian one, and by transferring knowledge among generations, the community was able to revive the commons after a long period of silence, caused by the centralised socio-political regime in the country. The community reinvented the forest commons' regime in the newly emerging context after 1990, through recalling cognitive development values as an inspiration, diversifying its life sustenance means beyond merely forest provisioning services, cooperating with conventional governance institutions at local level, and establishing a network of 256 forest users' associations. The latter represents a social innovation that has empowered the "nested enterprise" character of the forest commons in Albania, allowing for cross-scales interaction to take place, therefore strengthening the polycentricity pattern of forests' territorial governance.

On the other hand, the case shows also that the successes of the commons at the small scale cannot be transferred to the larger scale, unless (1) conventional forest governance embraces all of the dimensions of territorial governance and (2) scientific knowledge is transferred to the community in an exchange between experts and commoners. Mutual exchange of traditional and scientific knowledge is a crucial factor in overcoming the territorial scales mismatch in the governance of natural resources and socio-ecological interactions.

Finally, while the forest commons' regime embodies all of the territorial governance dimensions (with some being more prominent than the others), this is not the case for the forests conventional governance. Of course, the commons succeed (among others) because the conventional governance institutions have allowed this system to develop, even in the absence of the legal framework. Still, the non-common forests experience severe deforestation. The only way to revert this situation is through revising forest conventional governance to gain the territorial perspective, learn from the small-scale practices, embed on a knowledge transfer process, and employ a model of natural resources governance (as proposed in this chapter), where the relationship between supply and demand for ecosystem services is balanced and facilitated by governance institutions that cooperate across scales.

References

Agrawal A (2000) Small is beautiful, but is larger better? Forest management institutions in the Kumaon Himalaya, India. In: Gibson CC, McKean MA, Ostrom E (eds) People and forests: communities, institutions and governance. MIT Press, Cambridge, pp 57–86

Agrawal A (2003) Sustainable governance of common-pool resources: context, methods, and politics. Annu Rev Anthropol 32:243–262

Alcántara CHD (1998) Uses and abuses of the concept of governance. Int Soc Sci J 155:105–112

Alexander GS, Penalver EM (2012) An introduction to property theory. Cambridge University Press, New York

Araral E, Hartley K (2013) Polycentric governance for a new environmental regime: theoretical frontiers in policy reform and public administration. In International conference on public policy. Grenoble, 2013. IPPA (International Public Policy Association). http://archives.ippapublicpolicy.org/IMG/pdf/panel_46_s1_araral_hartley.pdf. Accessed 23 Apr 2018

Armitage D (2008) Governance and the commons in a multi-level world. Int J Commons, 2, pp 7–32. http://www.thecommonsjournal.org; URN:NBN:NL:UI:10-IJC-08002. Accessed 23 Apr 2018

Arnouts R, Van Der Zouwen M, Arts B (2012) Analysing governance modes and shifts — governance arrangements in Dutch nature policy. Forest Policy Econ 16:43–50

Beatley T (1994) Ethical land use: principles of policy and planning. Johns Hopkins University Press, Baltimore

Bertrand F, Richard E, Larrue C (2016) Climate problem and territorial governance: an overview of adaptation initiatives at the French regional level. In: Knieling JGF (ed) Climate adaptation governance in cities and regions: theoretical fundamentals and practical evidence, 1st edn. Wiley, Hoboken, NJ, pp 191–208

Blatter J (2004) 'From Spaces of Place' to 'Spaces of Flows'? Territorial and functional governance in cross-border regions in Europe and North America. Int J Urban Reg Res 28(3):530–548

Böhme K, Zillmer S, Toptsidou M, Holstein F (2015) Territorial governance and cohesion policy study. Directorate General for Internal Policies. http://www.europarl.europa.eu/studies. Accessed 23 Apr 2018

Braat LC, de Groot R (2012) The ecosystem services agenda: bridging the worlds of natural science and economics, conservation and development, and public and private policy. Ecosyst Serv 1(1):4–15

Brnkaláková S (2016) Adaptive management of mountain ecosystem services. Dissertation Thesis, Slovak University of Technology, Bratislava

Burkhard B, Muller F (2015) Indicators and quantification approaches. In: Grunewald K, Bastian O (eds) Ecosystem services - concept, methods and case studies. Springer, Berlin, pp 76–85

CEMAT (2010) Council of Europe Conference of Ministers responsible for Spatial/Regional Planning (CEMAT): basic texts 1970-2010Exts. Territory and Landscape No. 3. Council of Europe. https://rm.coe.int/16804895e4. Accessed 23 Apr 2018

Chan KMA et al (2011) Cultural services and non-use values. In: Kareiva P et al (eds) Natural capital: theory and practice of mapping ecosystem services, 1st edn. Oxford University Press, Oxford

CoR (2014) On the mid-term review of Europe 2020: Athens declaration - a territorial vision for growth and jobs. Athens: Committee of the Regions Committee of the Regions. http://cor.europa.eu/en/news/Documents/2210-athens-declaration-a5.pdf; http://cor.europa.eu/en/news/Pages/regions-cities-athens-declaration.aspx. Accessed 23 Apr 2018

Cox M (2014) Understanding large social-ecological systems: introducing the SESMAD project. Int J Commons 8(2):265–276

Cox M, Arnold G and Villamayor-Tomás S (2010) A review of design principles for community-based natural resource management. Ecol Soc, 15(4), p.38. http://www.ecologyandsociety.org/vol15/iss4/art38/. Accessed 23 April 2018

CSD (1999) ESDP European Spatial Development Perspective: towards balanced and sustainable development of the territory of the European Union. European Commission Committee on Spatial Development, Potsdam. http://ec.europa.eu/regional_policy/sources/docoffic/official/reports/pdf/sum_en.pdf. Accessed 24 Apr 2018

Davoudi S, Cowie P (2016) 'Good' territorial governance. In: Schmitt P, Well LV (eds) Territorial governance across Europe pathways, practices and prospects. Routledge, Oxon, pp 48–59

Davoudi S, Evans N, Governa F, Santangelo M (2008) Territorial governance in the making: approaches, methodologies, practices. B Asoc Geogr Esp 46:351–355

Derissen S, Quaas M, Baumgärtner S (2009) The relationship between resilience and sustainable development of ecological-economic systems. Working Paper Series in Economics, 146. https://www.econstor.eu/bitstream/10419/30245/1/61690682X.pdf. Accessed 24 April 2018

Driessen PPJ et al (2012) Towards a conceptual framework for the study of shifts in modes of environmental governance – experiences from the Netherlands. Env Pol Gov 22:143–160

European Commission (2009) Green paper on territorial cohesion – turning territorial diversity into strength. [Online] http://ec.europa.eu/regional_policy/archive/consultation/terco/consultation_en.htm. Accessed 23 Nov 2017

European Commission (2018) A credible enlargement perspective for and enhanced EU engagement with the Western Balkans. COM(2018) 65 final. European Commission European Commission, Strasbourg. Communication from the Commission to the European Parliament, the Council, the European Economic and Social Committee and the Committee of the Regions; https://ec.europa.eu/commission/sites/beta-political/files/communication-credible-enlargement-perspective-western-balkans_en.pdf. Accessed 24 Apr 2018

Faludi A (2012) Multi-level (territorial) governance: three criticisms. Plan Theory Pract 13(2):197–211

Faludi A (2016) Territorial governance challenging government. In: Schmitt P, Van Well L (eds) Territorial governance across Europe: pathways, practices and prospects. Routledge, Oxon

FAO (2015) Global Forest Resources Assessment 2015. [Online] FAO. http://www.fao.org/3/a-i4808e.pdf. Accessed 27 Aug 2017
FAO (2016) State of the World's Forests 2016 - forests and agriculture: land-use challenges and opportunities. FAO. Excerpt From: "State of the World's Forests 2016." iBooks
Finka M, Kluvánková-Oravská T (2010) Concept of polycentricity in current theory and practice of territorial governance. Prognostické práce 2(1):23–37
Finka M, Kluvánková T, Ondrejička V (2015) Concept of soft polycentric governance for fuzzy soft spaces as a challenge for central European peripheral spaces. In: Lang T, Henn S, Sgibnev W, Ehrlich K (eds) Understanding geographies of polarization and peripheralization perspectives from central and Eastern Europe and beyond. Palgrave Macmillan, New York, pp 309–322
Fleischman F, Loken B, Garcia-Lopez GA, Villamayor-Tomas S (2014/a) Evaluating the utility of common-pool resource theory for understanding forest governance and outcomes in Indonesia between 1965 and 2012. Int J Commons, 8(2), 304–336
Gjeçovi S (1925) Kanuni i Lekë Dukagjinit [The Kanun of Lekë Dukagjini]. Shtëpia Botuese "Kuvendi", Tiranë
GoA (1992) "Për pyjet dhe policinë e shërbimit pyjor" (law on forests and forests police) no. 7623. Qendra e Botimeve Zyrtare, Tiranë
GoA (2005) "Për Pyjet dhe Shërbimin Pyjor" [Law on forests and forestry service] No.9385. 04 May. p 1573
Goulder L, Kennedy D (2011) Interpreting and estimating the value of ecosystem services. In: Kareiva P et al (eds) Natural capital: theory and practice of mapping ecosystem services, 1st edn. Oxford University Press, Oxford
Grunewald K, Bastian O, Syrbe RU (2015) Space and time aspects of ES ecosystem services: concept, methods and case studies. Springer, Berlin
Grünwald A, Wende W (2015) Integrating the concept of ecosystem services into landscape planning. In: Grunewald K, Bastian O (eds) Ecosystem services, concept. Methods and case studies. Springer, Berlin, pp 165–172
Haines-Young R, Potschin M (2011) Common International Classification of Ecosystem Services (CICES): 2011 Update. [Online] European Environment Agency. Paper prepared for discussion at the expert meeting on ecosystem accounts organised by the UNSD, the EEA and the World Bank. https://cices.eu/content/uploads/sites/8/2009/11/CICES_Update_Nov2011.pdf. Accessed 22 Aug 2017
Haines-Young R, Potschin M (2012) Common International Classification of Ecosystem Services (CICES): consultation on Version 4. [Online] European Environment Agency. Report to the European Environment Agency; EEA Framework Contract No EEA/IEA/09/003. https://cices.eu/content/uploads/sites/8/2012/07/CICES-V43_Revised-Final_Report_29012013.pdf. Accessed 22 Aug 2017. Revised January 2013
Hajer M, Wagenaar H (2003) Introduction. In: Hajer MA, Hendrik Wagenaar H (eds) Deliberative policy analysis: understanding governance in the network society. Cambridge University Press, Cambridge
Halkier H (2012) Knowledge dynamics and policies for regional development: towards a new governance paradigm. Eur Plan Stud 20(11):1767–1784
Handmer JW and Dovers SR (1996) A typology of resilience: rethinking institutions for sustainable development. Organization Env, 9(4), p.482. http://oae.sagepub.com/content/9/4/482. Accessed 24 Apr 2018
Healey P (2006) Transforming governance: challenges of institutional adaptation and a new politics of space. Eur Plan Stud 14(3):299–320
Hooghe L, Marks G (2003) Unravelling the central state, but how? Types of multi-level governance. Am Polit Sci Rev 97(2):233–243
Hysing E (2009) From government to governance? A comparison of environmental governing in Swedish Forestry and Transport. Governance 22(4):647–672
INSTAT (2016) Fondi Pyjor sipas Pronesisë [Forest Fund as per Ownership]. [Online] http://www.instat.gov.al/al/themes/mjedisi.aspx. Accessed 24 Apr 2017

Kareiva P et al (eds) (2011) Natural capital, theory and practice of mapping ecosystem services. Oxford University Press, New York

Keating M (2009) Rescaling Europe. Perspect Eur Polit Soc 10(1):34–50

Kluvánková T, Gežík V (2016) Survival of commons? Institutions for robust forest social-ecological systems. [Online] Elsevier. https://doi.org/10.1016/j.jfe.2016.01.002. Accessed 2 Mar 2017

Kovač P, Tomaževič N, Leben A, Aristovnik A (2016) Reforming public administration in Slovenia: between theory and practice of good governance and good administration. Int J Public Policy 12(3/4/5/6):130–148

Kooiman J (ed) (1993) Modern governance: new government–society interactions. Sage, London

Kooiman J (ed) (2003) Governing as governance. Sage, London

Lange P (2017) Sustainability governance: exploring the potential of governance modes to promote sustainable development. Nomos

Li TM (2006) Neo-Liberal Strategies of Government through Community: the Social Development Program of the World Bank in Indonesia. International Law and Justice Working Papers, Global Administrative Law Series

Low B et al (1999) Human–ecosystem interactions: a dynamic integrated model. Ecol Econ 31:227–242

Luque S, Gonzalez-Redin J, Furst C (2017) Mapping forest ecosystem services. In: Burkhard B, Maes J (eds) Mapping ecosystem services. Pensoft Publishers, Sofia, pp 324–328

MA (2005a) Ecosystem and human Well-being synthesis. Island Press Millennium Ecosystem Assessment, Washington, DC

MA (2005b) Current state and trends assessment - chapter 21: forest systems. Island Press Millennium Ecosystem Assessment, Washington, DC

Maes J et al (2012) Mapping ecosystem services for policy support and decision making in the European Union. Ecosyst Serv 1:31–39

Marsh WM (2010) Landscape planning environmental applications, Fifth edn. Wiley, Danvers

Meadowcroft J, Langhelle O, Ruud A (2012) Governance, democracy and sustainable development: moving beyond the impasse. In: Meadowcroft J, Langhelle O, Ruud A (eds) Governance, democracy and sustainable development: moving beyond the impasse. Edward Elgar Publishing Limited, Glos, pp 1–13

Merino L (2015) Reflections on the governance of the global commons. Commons Digest 17:10–13 Publication of the International Association for the Study of the Commons

Ministria e Ekonomisë Kombëtare (1930) Ligja e Pyjeve e Kullosave [Law on forests and pastures]. Botime të Ministrisë së Ekonomisë Kombëtare, Shtypshkronja "Tirana", Tiranë, Albania

Moran E (2005) Human-environment interactions in forest ecosystems: an introduction. In: Moran EF, Ostrom E (eds) Seeing the forest and the trees: human-environment interactions in forest ecosystems. MIT Press, Cambridge

Muharremaj V (2003) Pyjet dhe Kullotat [Forests and pastures]. In: Agolli S, Përmeti M (eds) Historia e Bujqësisë dhe Agroindustrisë Shqiptare. KEA, Tiranë, Albania, pp 243–295

Muharremaj V et al (2009) Studim mbi Kuadrin Ligjor për Pyjet dhe Kullotat Komunale në Shqipëri [A study on the legal framework of communal forests and pastures in Albania]. SNV-Sida, Tiranë

Nagendra H, Ostrom E (2012) Polycentric governance of multifunctional forested landscapes. Int J Commons 6(2):104–133

National Agency of Protected Areas (2015) Protected areas [Zonat e Mbrojtura]. [Online]. http://akzm.gov.al/index.php?option=com_k2&view=itemlist&layout=category&task=category&id=5&Itemid=289⟨=en. Accessed 17 Mar 2017

OECD (2001) OECD territorial outlook. Paris: OECD. http://www.keepeek.com/Digital-Asset-Management/oecd/urban-rural-and-regional-development/oecd-territorial-outlook_9789264189911-en#page1. Accessed 1 Dec 2017

Olson M (1965/2002) The logic of collective action - public goods and the theory of groups, 20th edn. Harvard University Press, Cambridge

Ostrom E (1990) Governing the commons, the evolution of institutions for collective action. Cambridge University Press, New York

Ostrom E (1999) Design principles and threats to sustainable organisations that manage commons. [Online] https://www.researchgate.net/publication/42766234_Design_Principles_and_Threats_to_Sustainable_Organizations_that_Manage_Commons. Accessed 16 Aug 2017

Ostrom E (2003) How types of goods and property rights jointly affect collective action. J Theor Policies 15(3):239–270. https://doi.org/10.1177/0951692803015003002

Ostrom E (2009a) Beyond markets and states: polycentric governance of complex economic systems. [Online] http://www.nobelprize.org/nobel_prizes/economic-sciences/laureates/2009/ostrom_lecture.pdf. Accessed 15 August 2017. Prize Lecture

Ostrom E (2009b) Design principles of robust property-rights institutions: what have we learned? Lincoln Institute of Land Policy, Cambridge, MA

Ostrom V, Tiebout CM Warren R (1961) The organization of government in metropolitan areas: theoretical inquiry. Am Polit Sci Rev, 55(4), pp.831–842. http://www.jstor.org/stable/1952530. Accessed 24 Apr 2018

Palahi M, Mavsar R, Gracia C, Birot Y (2008) Mediterranean forests under focus. Int For Rev 10(4):676–688

Pejovich S (1990) The economics of property rights: towards a theory of comparative systems, International studies in economics and econometrics. SpringerScience+Business Media, Dordrecht

Perrings C (2006) Resilience and sustainable development. Env Dev Econ 11:417–427

Piattoni S (2010) The theory of multi-level governance: conceptual, empirical, and normative challenges. Oxford University, Oxford

Rametsteiner E (2009) Governance concepts and their application in forest policy initiatives from global to local levels. Small-scale Forest 8(2):143–158

Raymond L (2003) Private rights in public resources, equity and property allocation in market-based environmental policy. RFF Press, New York

Raymond R (2008) Agreements and controversies around the notion of territorial governance: a bibliographic review of a fashionable notion. Int J Sustain Dev 11(2/3/4):115–137

Rhodes RAW (1996) The new governance: governing without government. Politicol Stiidia XLIV:652–667

Rivolin UJ (2010) Spatial units for EU territorial governance: findings from a study on north-western Italy. Eur Plann Stud 18(2):299–316

Satake A, Rudel TK, Onuma A (2008) Scale mismatches and their ecological and economic effects on landscapes: a spatially explicit model. Glob Environ Challenge 18(4):768–775

Schiappacasse P, Müller B (2018) Identifying gaps and opportunities for research on urban and regional resilience—highlighting the advantages of research cooperation. In: Müller B, Shimizu H (eds) Towards the implementation of the new urban agenda. Springer, Berlin, pp 259–270

Schmitt P, Well LV (eds) (2016a) Territorial governance across Europe: pathways practices and prospects. Routledge, Oxon

Schmitt P, Well LV (2016b) Revisiting territorial governance: twenty empirically informed components. In: Schmitt P, Well LV (eds) Territorial governance across Europe: pathways, practices and prospects. Routledge, Oxon, pp 221–237

Schmitt P et al (2013) ESPON TANGO – Territorial approaches for new governance. Scientific report. ESPON & Nordregio ESPON, Luxembourg

Shoeibi S, Bürger-Arndt R, Hubo C (2015) Forest ecosystem services - the state of European Policies and Scientific Knowledge. Master Thesis, OmniScriptum GmbH&Co. KG., Saarbrücken

Shutina D, Toto R (2010) Parathenie. In: Shutina D, Toto R (eds) Politikëndjekës apo Politiëbërës: Alternativa mbi zhvillimin urban, manaxhimin e territorit dhe të mjedisit [Policyfollowers or Policymakers: Alternatives on urban development, territorial and environmental management], 1st edn. Co-PLAN and IKZH_POLIS, Tiranë, pp 9–14

Shutina D, Toto R (2017) Unraveling territorial governance in Albania – a policy decision or an intuitive approach? In Territorial governance and spatial planning in the Western Balkan Region. Torino, 2017, Politecnico di Torino
Stead D (2013) Dimensions of territorial governance. Plann Theory Pract 14(1):142–147
Stead D (2014) The rise of territorial governance in European policy. Eur Plann Stud 22(7):1368–1383
Stead D, Pálné Kovács I (2016) Shifts in governance and government across Europe. In: Schmitt P, Well LV (eds) Territorial governance across Europe: pathways, practices and prospects. Routledge, Oxon, pp 21–35
Stoker G (1998) Governance as theory: five propositions. Int Soc Sci J 50(155):17–28
Toto R (2018) Ecosystem-based governance of forest commons - the case of Shkumbini River basin. IDAUP Ferrara University, POLIS University, Tiranë, Albania
UN-Habitat (2002) The global campaign on urban governance, Concept paper, 2nd ed. UN-Habitat. https://unhabitat.org/books/global-campaign-on-urban-governance-the/. Accessed 24 Apr 2018
United Nations (1987) Report of the world commission on environment and development: our common future. http://www.un-documents.net/our-common-future.pdf. Accessed 5 Aug 2017
United Nations (2014) The value of forests, payments for ecosystem services in Green Economy. ECE/TIM/SP/34. Geneva, United Nations
Van Oel R, Krol MS, Hoekstra A (2009) A river basin as a common-pool resource: a case study for the Jaguaribe basin in the semi-arid Northeast of Brazil. Int J River Basin Manage 7(4):345–353
Well LV, Schmitt P (2015) Understanding territorial governance: conceptual and practical implications. Eur Reg 21(4):209–221
Well LV, Schmitt P (2016) Territorial governance across Europe: setting the stage. In: Well LV, Schmitt P (eds) Territorial governance across Europe: pathways, practices and prospects. Routledge, Oxon, pp 3–20
WWF (2017) Forest habitat. https://www.worldwildlife.org/habitats/forest-habitat. Accessed 25 Aug 2017
Young OR (2013) On environmental governance: sustainability, efficiency, and equity, Series: On politics, 2nd edn. Routledge, Oxon

Chapter 8
The Role of Leadership in the Current Contexts of Central European Planning Culture

Matej Jaššo and Maroš Finka

8.1 Core Ideas

> *"The character of planning culture is determined by the value system, which is mediated by und mirrored in the acting and communication".*
>
> (*Finka 2006*)

> *"Understanding of cultural dynamics, the ability to develop a clear vision of the desired planning culture and its active fostering, should be an essential competence of any leader".*
>
> (*Jaššo 2011*)

8.2 Introduction

Planning culture has recently become one of the most intensively examined soft factors in spatial planning, being both the direct object of research focused on planning culture or indirectly creating the framework for understanding particular processes in European spatial development or framing the research on planning instruments particularly linked to specific development problems. Our contribution refers to the outputs from the research projects under the participation of its authors, especially research for European spatial planning handbooks series (ARL Hannover 2000–2006) including the desk research on the development of national planning systems within Central Europe and their comparison externalized within

M. Jaššo (✉) · M. Finka
Spectra Centre of Excellence of the EU, Slovak University of Technology in Bratislava, Bratislava, Slovakia
e-mail: matej.jasso@stuba.sk; maros.finka@stuba.sk

© Springer International Publishing AG, part of Springer Nature 2019

M. Finka et al. (eds.), *The Role of Public Sector in Local Economic and Territorial Development*, EAI/Springer Innovations in Communication and Computing, https://doi.org/10.1007/978-3-319-93575-1_8

analytical work within the European projects BROWNTRANS, BRIBAST, CIRCUSE, and LUMAT and workshop Planning Culture in the Central and Eastern European Countries (29th June 2005) in Bratislava. The desk research was completed by the questionnaire survey (Roch et al. 2008; ARL EAG Central and Eastern Europe). Special attention has been paid to the role of the leadership in fostering, steering, and developing desired planning culture.

Recent societal development requires radical shift from old regulative planning approaches based on more or less rational-predictive models. Globalization and fragmentation of society as well as the multimodality of various social and economic processes have yielded a new situation: planning in general became rather managerial process of fostering, steering, and influencing of plethora of factors than prescription of future development following certain professional vision of optimal target quality defined by a planner. Space is no more a "neutral" category or a "container" for social and economic processes as it was considered till the 1980s, but it is rather the result of social relations among people (Knieling and Othengrafen 2009). Desire for optimal outcome predefined by professionals has been replaced by achieving compromise and satisfaction of the whole spectrum of the stakeholders in spatial development process. To start theoretical dispute focused on background, roots, methodology, models, and research on planning culture and the role of leadership in planning in general, it is necessary to bring some initial notions on broader contexts and frameworks, from which planning culture gains its origin, strength, and vitality.

One of the most significant changes in spatial development in recent decades is concerning the role of the planners. It has been continuously modified from the position of the designers of the bright future toward the role of mediators and leaders within the processes of spatial development. The leadership qualities are essential and in many complicated, multitasking planning processes they are becoming the focus of whole planning activities. (Post)modern spatial planning has to deal intensively with immaterial and qualitative variables and phenomena like community, identity, place, social behaviour, or human values. Recent global development has highlighted the need for special attention dedicated to these issues especially in the context of territorial fixing of the mobile capital. Highly profiled planning cultures based on the active leadership generating typical set of principles, values, and approaches play an important role in the spatial development of any territory. Planning culture and the quality of leadership related to planning significantly shape not only territorial and spatial processes, but also the dynamics of the entire society. Analysis of the past has often shown that especially planning cultures based upon the authoritarian and bureaucratic leadership failed in the process of dealing with challenges of high complexity. Planning is an integral part of the societal reflection and interactions with the environment—uniqueness of environment generates unique reflections, especially in terms of culture and leadership. Planning culture could be deemed as specific, unique as well as typical, non-casual set of approaches to the factors playing principal roles in the process of spatial development. It can be considered as one of the softest dimensions of spatial planning (an important non-proprietary part of social and territorial capital). Planning culture is a collective asset without clear boundaries but might be significantly influenced by individual

Table 8.1 Relation of different planning paradigms toward the role of the planner and leadership

Paradigm	Philosophical background	Role of the planner	Leadership
Rational-predictive planning	Positivism, marxism	Social engineer, expert	Very limited
Progressive incrementalism	Liberalism, pragmatism, critical theory	Coordinator, expert, decision-maker	Limited, but pragmatic
Advocacy planning	Neoliberalism, pragmatism	Advocate, expert	Developed in certain fields
Participative (communicative) planning	Post-structuralism, theory of communicative action	Facilitator, narrator, mediator leader	Essential, but exposed to democratic competition

actions—mature leadership is always a symbol of certain culture. Last but not least—there is highly evocative and motivational potential of leadership. Many planning processes of recent past have failed on the general feeling of tiredness and they did burn out. Authentic leadership having roots in balanced, mature, and democratic planning culture might be an answer also in these cases (see more in Table 8.1).

8.3 Contexts and Frameworks of the Development of Planning Cultures in Europe

Geopolitical and Historical Background

Central Europe has been continuously the centre of attention of various academic disputes, political discourse as well as the overall dialogue on Europe. This territory has been largely volatile and instable during recent centuries and became a place of many territorial shifts, redesigns of boundaries as well as the birth and rebirth of plethora of identities (ethnic, cultural, political). Semiotics of the term "Central Europe" is very wide and manifold. Once a cover sheet for many *clichés* (Austro-Hungarian Empire, "Eastern Europe", "communist block") is today a platform for various forms of regional cooperation (CENTROPE, Twin City, Czech-Slovak cooperation, etc.). It has been tackled in many essential politological and sociological texts, often putting this area in the twilight of doubt and ambiguous circumstances. Timothy Gordon Ash in his famous text "Does central Europe exist?" states that according to former Austrian politician F. Sinowatz the word Central Europe was "almost not permitted" and German word "Mitropa" was predominantly linked to the name of the dining cars of the Deutsche Reichsbahn (see Ash 1986). Milan Kundera (1984) in his highly emotional essay "The Tragedy of Central Europe" (published in West in the time of nadir of overall societal and economical decay in former Czechoslovakia) questions the communism as the "evil imported from the East" within this highly sensitive territory and protests against the old labels routinely evoking the "East-European" semiotics (e.g. slightly derogative label "Ostblock") in the Central European discourse. Even in 1990s, in the times when the name "Central Europe" had been officially mentioned in the parliament speech of British Queen Elisabeth II and the British Foreign Ministry established "Central European Department" (Ash 1999), many politicians and decision-makers privately used former and obsolete "East Block" rhetoric figures.

According to Bělohradský (2013), there are three most important historical prerequisites of forming the (post)modern interpretation of Central European identity and culture: (1) highly instructive demise of Austro-Hungarian Empire as a supranational state, (2) Viennese modernism with its fragmentation of tradition, and (3) "turn to language" which anticipated later post-modern movements and prevailing historical consciousness of Western Europe in 1970s–1980s condemning scandalous fact that cities like Prague, Budapest, Bratislava, or Ljubljana are (seemingly) permanently missed in European discourse (see Bělohradský 2013, p. 302).

Central Europe was always a prime example of all the progress, achievements as well as later disappointments and setbacks related to the era of organized modernity. V. Bělohradský (2013) interprets Central Europe during the period of modernity as "a place where people underwent an existential experience of great historical importance for modern Europe, an experience whose fateful meaning Central European literature and philosophy sensed with an exceptional perspicacity" (p. 303). Whole Central European philosophy and literature analysed "the absurdity of the impersonal language of legality" (Bělohradský 2013) having its roots in Austro-Hungarian empire and having been drawn to perfection by communist regimes. Central Europe of the twentieth century has been witnessing extraordinary examples of authentic political leadership (T.G.Masaryk and M.R.Štefánik in 1918, L.Walesa in 1981, V.Havel in 1989), tragic episodes when positive leadership abruptly vanished and the outcome was grim (A.Dubček in 1968, I.Nagy in 1956) as well as the examples of misused leadership and its transformation into the ruthless authoritarian policy against own nation (G. Husák or W.Jaruzelski in 1970s).

Although before 1989 the Central Europe was essentially split apart into two parts and the process of development of democratic planning culture was limited to countries belonging to western European territory, the traditional cultural roots embedded in the Habsburg monarchy followed by the progressive functionalist movement between two World Wars, residual power to sustain on rational principles in territorial planning was so strong, that further development of the territorial planning instruments did not tear apart from the western democracies in this space. After all, rational and predictive models of planning were the common paradigmatic platform for whole Europe till the end of 1970s (although being more ideologically pronounced in countries under communist regimes). Global shift to more cooperative and participative ways of planning in late 1980s exposed unsustainability of pure rationalistic models of planning, especially those which were politically and ideologically backed by old, obsolete, and burnt out communist ideology. The first attempts to form a civic society in Slovakia, back in the late 1980s, were often motivated and driven by environmental issues—e.g. non-formal initiative "Bratislava nahlas" [Bratislava Out Loud] (Budaj et al. 1987). This initiative was a primary example of the political and societal leadership born in the civic sector, which has been confirmed 2 years later, during the so-called velvet revolution in Czechoslovakia. Roots of participative, discourse-based planning culture as well as civic leadership in Slovakia might be traced in grim late 1980s.

In terms of spatial development, the situation on the other edge of the iron curtain was also far from optimal: the vital functions of the cities and regions located there were also under-nurtured: Burgenland, Lower Austria, and peripheral parts of Bavaria were rather low-key based and peripherally placed places of the democratic Europe. In 1990s, the process of geopolitical changes in Europe culminated by re-unification of Germany and political shifts in Central and East European countries and these events have delivered completely new spatial, social, and economic realm in Central Europe (Jaššo 2007). Interface between former West and East (represented, e.g. in Vienna-Bratislava cooperation) is a place, where the problems as well as the opportunities after the removal of iron curtain had appeared in utmost visible way (for more details, see, e.g. Jaššo 2007, 2008a; Finka 2005; Schimak 2003; Hrdina et al. 2002). We were witnessing two major trends: the lowering of the importance of national barriers with the forthcoming integration of national, regional, and local economies on the one hand, and the emergence of new development poles in the form of functional regions that cut across administrative borders on the other hand (Finka 2005; Jaššo 2007). Immediate information transfer hugely accelerated municipal development. Patterns of municipal competition and cooperation started to be externalized in the spatial practice. Initially, the democratic decentralization and revival of common tradition had substantial yield. Many regional initiatives of cooperation delivered promising results (Jordes+, Interreg IIIA,B) but later their ethos somehow faded away (CENTROPE, Twin City). Years after millennium brought also some portion of disappointment and several setbacks, related often to the dominance of neoliberal solutions even in the spots where the more structurally sensitive approaches were necessary (see, e.g. Finka and Jaššo 2010). The subsequent global economic crisis hit the Central Europe in moderate scale. Maybe the similarities in planning cultures (derived from overall cultural similarities) were among the factors, which made overcoming of these barriers and troubles easier. Several planning aspects of the CENTROPE initiative (Central European Region consisting from cities and regions from Austria, Czech Republic, Hungary, Slovakia) seem to revive and reflect the planning culture in the regional and urban scale of planning and spatial development—see, e.g. term "cultural imaginaries" (elaborated more in Suitner 2015).

The EU Approximation Processes and Enlargement After Millennium

The EU enlargement (especially the phase after millennium) and integration have belonged to the most prominent processes running in the European territory at the turn of the centuries. Cities and regions are getting closer to each other in terms of cooperation as well as competition. European integration could be seen as both—parallel phenomenon to the economic globalization on European level and as a response to the challenges resulting from the globalization processes. Unlike many spontaneous aspects of globalization, European integration process represents to big extend institutionalized process coordinated and controlled by administrative and legislative measures. In many fields of societal life (economy, environmental quality, transport, education, etc.), the integration was connected with fundamental influence of joint policies and the EU legislation on national states catalysing approximation processes in the institutional arrangements of the management processes in respective field. Integrative management of spatial development has left out of the main stream of these processes being closely connected with the territorial sovereignty across the national states, regions, and municipalities and territorial identity as a special kind of phenomenon, which forms throughout historical and territorial socialization (Raagmaa 2002). Natural pressure on the legal and factual approximation in spatial development management, including planning systems, represents the exposure of the territorial subjects (municipalities, regions, states) to the competition on the common European market, where they have to compete for investors, working places, inhabitants on one side, and their growing exposure to the global external shocks (e.g. effects of climate change, economic crises) requiring cooperation and joint strategies to lower their negative effects. The need to react flexibly to new challenges, dynamic changes in the framework preconditions, and the necessity to follow integrative strategies for more and more complex socio-ecosystems of the cities and regions on the one hand and long-term development sustainability requirements on the other hand catalyse the approximation processes of two basic approaches in spatial development management—ex ante and ex post planning in relation to the planning demand.

Growing Complexity, Fuzziness of the Objects of Planning, and Individual/Spatial Activities

Growth of the complexity of the cities and regions as the objects of planning is followed by the process of their fuzzification. The administrative borders of the cities, regions, and national states lost their importance; the physical as well as functional borders disappear in the mish-mash of complex functional ties realized by the flows of energies, materials, and peoples enjoying the freedom of growing mobility and individual decision-making. The management of spatial development process is embedded in this environment of uncertainty given by difficult predictable multi-actors decision-making in fuzzy spaces across administrative territorial units and across their hierarchies. The fuzziness relates not only to physical or different cultural spaces, but also to social and cultural background of the holders of the decision-making rights and other stakeholders in spatial development process.

Growing Sensitivity to the Soft Factors

Human value system is in process of permanent and continuous change. The shift in the value model has never been more obvious and abrupt as nowadays. Beginning from the 1990s, phenomena like community, identity, place, social cohesion, or human values gained keen attention of the spatial planners and all those who are dealing with spatial planning on various academic, scientific, educational, or expertise levels. Sustainability of any societal as well as spatial system is determined by its inner cohesion. Cohesion secures stability of the system, its endurance and composure (Finka and Jaššo 2010). Unique and delicate combination of mutual interaction of various systems within the territory (economic, social, ecological, cultural) generates high demands on the sustainability. Social cohesion and quality of the social structures plays an important role in this process. Balanced and durable social cohesion is utmost important for the competitiveness of any territorial subject and effective allocation of its social and cultural capital. Culture (including planning culture) is becoming the vital bond between past and future, between experience and theoretical knowledge, and between the success and failure.

Growing Uncertainty in the Decision-Making

Exact, normative, and sometimes rigid decision-making was once the most fundamental hallmark of all authoritarian planning cultures based on the rational-predictive models. Even the western democratic countries abandoned this tradition relatively late (in 1980s). Transformation of the internal as well as external frameworks and conditions in 1980–1990s has had huge impact on planning. And the old Kuhn hypothesis on paradigm shift has been revived once again—change must be made by those living and acting outside the prevailing paradigm (Kuhn 1970). This is reflected also in the Schein's (1992) assertion that leader must somehow step out from the culture in which he/she has been born and grown up. Planning became rather contingent in its nature (Keim et al. 2002) without any warranty of the successful outcome. This brought new situation—planning and decision-making in planning process is almost always "judgement under certain degree of uncertainty" (see, e.g. Jaššo 2008b, c, 2011; Finka and Kluvánková 2015). Many statistical, stochastic, and econometric models helped assess the risk/benefit ratio, but the definitive methodological and predictive answers remained missing in most cases. Ambiguity became ever present part of planning game: it is sometimes difficult to analyse the diffuse and permanently changing motivational patterns of the stakeholders who are not steady and legible anymore. They are creating various ad hoc alliances, changing their behavioural patterns and needs in continuous way. Postmodern city is far away from the old models from Chicago school of sociology and human ecology models in the early years of modernity.

General and Deep Changes in Spatial Planning

Planning is not anymore a prescriptive set of rules, standards, and tools applied to particular territory. The role of the planner has been substantially modified: planner is no more an "architect of the future", creator, author, or mentor, but is rather coordinator and manager of the spatial development processes determined by multi-actors' decision-making. His/her tasks have been shifted from design to mediation and negotiation, from the definition of the target quality to process management. Overall planning paradigm has gone through huge changes—we were witnessing the shift from the system theories ("comprehensive planning") having their roots back in 1950s toward the "incrementalism" of the 1970s and later to cooperative planning of 1990s (for more details, see, e.g. Jaššo 2011). Changes of planning instrumentarium are covering various shifts from hard to soft instruments, from plans to process interventions. The focus on processes and phenomena beyond the traditional classification schemes and patterns replaced the focus of measurement, classification, and top-down intervention. Various bottom-up initiatives in 1990s brought the request for "communication turn", which is reflected in continuous proliferation of consensus and cooperation-based behavioural patterns and proceedings. Each democratic planning culture is based on the involvement of the non-professionals and has numerous feedback and controlling mechanisms. We see the synergy of new and traditional or formal approaches (e.g. sustainability paradigm—ex ante and problem-oriented participative planning ex post). Redefinition of the relations between the planners and the external partners is now commonly regarded as a more intelligent approach (Rode 2006). These new qualities/requirements/challenges are common denominators for new emerging planning cultures.

8.4 Planning Culture Concepts and Theories

Smooth, successful, and genuine spatial development requires the cultural compatibility and continuity. Integration of different values, basic assumptions, and beliefs into a coherent spatial concept is a necessity and ultimate challenge for spatial planners. Forester's concept "making sense together" has been completed by Healey's addition "while living differently" (Mäntysalo 2005). Efficiency of consensus-oriented planning culture or cooperative planning effort is strongly dependent on the degree of trust and tolerance. Initial consideration on the nature and scope of the planning culture, its fundamental layers and dimensions, need for scientific research, and further surveys of planning culture as well as the typical traits of Central European planning culture will be in the spotlight.

Planning culture is embedded in the society and its culture, as well as within the physical territory. According to Knieling and Othengrafen (2009), the term planning culture refers to the different planning systems and traditions, institutional arrangements of spatial development, and the broader cultural context of planning. We have to bear in mind that planning culture as an important phenomenon significantly influencing not only spatial development, but the society as a whole. It could be deemed as specific, unique as well as typical, non-casual set of approaches (based on underlying values) toward the factors playing fundamental roles in the process of spatial development within certain territory (see, e.g. Jaššo 2008b, c, 2011). Culture is generally perceived as a proven and elaborated way how to deal with the unknown, based on the collectively shared experience. Culture is always multidimensional and multifocal. It is deeply bound to collective perception of the people and is reflecting the ways and patterns how things have been implemented and evaluated and what had been proven right and what wrong in long-term temporal and spatial scales. Every culture is interpretative and has various connotations—knowledge of the interpretation codes distinct the insiders from outsiders. Sense of belonging to the certain culture is based upon the ability to read and interpret these codes. Hofstede (2001) defines culture as "collective programming of the mind, consisting of two poles: human nature and individual personality". Each culture has diffuse borders, and this is essentially valid also for planning culture. Planning culture might be interpreted as one of the cultural subsystems (similarly like corporate culture, culture of various professions, regional culture) existing within main cultural framework in the society. Underlying values, thoughts, and principles gain a spatial reflection within certain territory. Interaction and accumulation of data, knowledge, and experience are engines of every planning culture (Jaššo 2011).

Many definitions of the essence what planning culture means have been elaborated in recent decades. Deep, proper, and thorough methodological and theoretical review is quite challenging task. Fürst (2007) complains about missing comprehensible methodology, despite the heavy attention which has been paid to planning culture research in recent years. We tried to differentiate the following approaches:

Operational Definitions

Operational definitions try to capture the essence of the phenomenon of planning culture. Taking into consideration that planning culture is able to observe in laboratory conditions, and it is lived outside with many uncontrolled interfering variables, these approaches are rather descriptive. One of the first survey on planning culture, conducted in early 1990s by Keller et al. (in Sanyal 2005, p. 30) concentrates on:

- The tasks and subject matter of planning,
- The organization, structures, and legal frameworks in which planning practices are embedded,
- The background assumptions, values, and general orientation that guide planning endeavours.

Rode (2006) simply states that "different social contexts and values are ultimately generating particular planning approach", without further analysis of these contexts. Dollinger (In Fürst 2007) refers to "the ways how the decision-makers are dealing with planning processes". These definitions are suitable enough for spatial practice but we feel that further precision is necessary. Diller (in Umfrage zur Planungskultur 2007) sees the scope of the planning culture in the self-perception of actors being involved in spatial planning, their various interactions, and collaborative efforts.

Behavioural Definitions

This type of definitions is based upon the assumption that definition must be based on observable outputs (e.g. behavioural patterns). Friedmann (2005) defines planning culture as "the ways, both formal and informal, that spatial planning in a given-multi-national region, country or city is conceived, institutionalized and enacted". Ludwig (In Fürst 2007) defines planning culture as "the patterns of behavioural interaction between planners and stakeholders". According to Beutl (in Umfrage zur Planungskultur 2007), planning culture is the mode and ways of coordinating the spatial behaviour of various actors, not only planners and stakeholders. In this approach, planning culture is co-defined by the planning system (formal instruments), values and norms (non-formal instruments), and the individual traits and characteristics of the actors (interests, motivation, abilities). All these definitions attempt to include into scope of the planning culture predominantly visible, observable factors although we admit that above mentioned Beutl's (in Umfrage zur Planungskultur 2007) definition goes deeper.

Holistic Definitions

Most of the authors do agree that each culture has visible as well as invisible, "deeper" content, and this should be reflected also in the definitions of planning culture. Geertz (1973) defines culture as "a system of inherited conceptions expressed in symbolic forms by means of which men communicate, perpetuate, and develop their knowledge about and attitudes toward life". This is obvious when analysing very well-known general model of culture provided by Gullestrup (Gullestrup 2007, Knieling and Othengrafen 2009, p. 6). Gullestrup's model of planning culture consists of the vertical and horizontal dimension. Horizontal dimension integrates segments demonstrated in the same level: processing/technological segment, distribution segment, social segment, managerial and decision-making segment, conveyance segment, integration segment, identity-creating segment, and security-segment. Adjusted to planning culture theory, processing segment covers work with data and statistics, identity segments cover the role, prestige, and self-image of spatial planners, managerial/decision-making segments represent the way, and the manner how planning decision are being made and which leadership style is preferred, etc. Vertical dimension includes cultural layers which are hierarchically established (some of them are clearly visible, some are semi-hidden or completely hidden): manifested symbols, immediately perceptible layers, barely perceptible layers, formalized norms and rules, core culture layers, non-perceptible layers, basic values, and the fundamental concept of universe. This has been reflected in many other concepts of various cultural subsystems, e.g. Schein's (1992) concept of corporate cultures which integrates basic assumptions, underlying values, and material artefacts into one coherent concept. His (Schein 1992) concept is thoroughly examining and explaining mutual relationship between leadership and culture, making both field a part of one coherent societal structural framework. Anyway, somehow similar model (in the field of planning culture), based upon the visible, manifested expressions as well as on the non-visible underlying assumptions has been adopted from the recent work of Knieling and Othengrafen (2009) and introduced by Steinhauer (2011): planning culture is a synthesis of societal environment (self-conception of planning, societal models, concept of justice), planning environment (planning semiotics, instruments and procedures, tradition, scopes and range of spatial planning, economic structure), and planning artefacts (urban design, plans, data, laws, instrument, and procedures). Knieling (2007) offered a model of planning culture integrating macro level (societal significance of planning, basic value layers, patterns, norms), mezzo level (objectives and principles of planning, planning semiotics and semantics, political, administrative, economic and organizational structures) and micro level (actors, instruments, strategies, procedures, communication, participation). While macro level is not always visible but is necessary to deliver more context and framework, micro level is tangible, obvious and concrete and provides more focus and action. Planning culture in more narrow sense might also be considered as a subculture of urbanists, planners, geographers, developers, and other actors involved in planning processes, which are producing and sharing cognitive frames, practices, knowledge, beliefs, norms and rules, values and codes (Knieling and Othengrafen 2009; Hansen 2003). Faludi (In Fürst 2007) emphasizes on the ethical dimension, presenting planning culture as "the collective ethos and dominant attitude of planners regarding the appropriate role of the state, market forces, and civil society in influencing social outcomes". Keim et al. (2002) defines planning culture as a "repertoire for planning practice", consisting from "the underlying planning assumptions, leading ideas and images, rationalities bound to consensus building and conflict management, norms and rules" on the one hand and the "planning styles" incorporating the utilized knowledge and forms of cooperation on the other hand. Regional specific culture, local tradition as well as the professional "corporate culture" of the planning bodies and institutions are significantly influencing the planning culture. Holistic definitions usually take into the consideration the leadership, which is one of the crucial elements of every planning culture.

8.5 Urban Semiotics and Planning Culture

Urban (spatial) semiotics is a part of urban theory approaching city as "text based on the grammar of spatial patterns and structures generating the meaning" (Jachna 2004). Urban and landscape shapes bear various significance and their meanings are negotiated through societal discourse delivering many connotations. There has been articulated the need for "socio-semiotic approach" (Gottdiener and Lagopoulos 1986) which would go far beyond pure physical/spatial parameters and involves "mythological epistemology"—study of ideologies and narratives which are symbolically recorded in urban structures as well as in planning documents. Here we find significant overlapping with the field of planning culture.

City is the focus of ideological and political discourse with all dominant philosophical themes—the truth, the justice, the power—all this interpreted in material and immaterial symbols. Gottdiener (in Pipkin et al. 1983, p. 101) defines urban semiotics as an "articulation of ideologies with the help of urban space". Public space is governed by "polylogy" (Bělohradský 2013)—multiplicity of vocabularies, canons, and languages, which are not ordered hierarchically and none of them does possess the hegemony. According to M. Castells (in Pipkin et al. 1983, p. 68), "urban space is not a written text, but rather a projection screen for changing social reality". Hence, planning culture is a "macro-realm" for those realities. Voulteenaho et al. (2012, p. 6) suggests "strategic textualization of the city by planners, politics and decision makers". For M. Choay (in Pipkin et al. 1983), the analysis of planning culture, especially the language of spatial practice, plans, and documents is the key to urban semiotics. Recent processes of re-semantization and de-semantization of urban/rural spaces are obviously reflecting the shifts in planning cultures. While modernity brought uniform spatial language and initiated the process of certain de-semantization ("spatial *clichés*"), the post-industrial society and post-modern counter currents in architecture and urban design inflicted huge process of re-semantization, which is multifocal and fragmented into many pieces. Semiotic dimension of planning culture has been emphasized by Žigrai (2005), who concludes that planning culture is "an ultimate reflection of place identity and regional background based on the principles of proximity, similarity, mutual influence, principle of *gestalt*," etc. In this point of view, planning culture is "meta-scientific" subject, located on the interface of history, culture, social, regional, and spatial sciences. It is a kind of superstructure integrating impacts of historical development (e.g. real estate reforms, political events), general shifts in paradigm during past years (e.g. sustainability tackling its mental dimension—for more details on the framework of sustainability and planning culture, see, e.g. Kühn and Moss 1998), or deliberate external interventions (e.g. legal harmonization).

In semiotics, each "text" (or metatext, to be more precise) has its symbolical partiture. Symbols are carriers of meaning, encapsulating high complexity into single, repeatable signs. This repertoire of symbols is giving a meaning to certain metatext and makes it intelligible for those who are familiar with them. Symbols and their interpretation are unique and make planning culture distinctive. Internal relations among symbols, their hierarchization and order, and the freedom of their interpretation may characterize the whole system more than anything else. As each planning culture is based upon its semiotic level, we can distinguish the following group of factors (Jaššo 2011):

Material, Urban, and Architectural Symbols

This is the most visible substance of planning culture and includes all essential architectural, urban, and visual artefacts located within certain territory. These symbols are exactly spatially and topologically defined which makes them chronicles of the previous development of planning culture. Architectural design, urban design as well as type of landscape influenced by past planning activities are the most fundamental examples of this kind of symbols. These signs reflect all the invisible parts of planning culture: values, underlying assumptions, collective ethos and moral principles, cultural aspects, etc. Their full and sound interpretation is possible only within these frames.

Spatial Configuration and Patterns on Macro Level

These symbols are not obvious and are observable only in macro level because they are not exposed to direct perception. Qualified reader must be able to distinguish them in more abstract way, e.g. in maps, planning document and strategies, aerial views, etc. Despite having their direct physical impacts, they are reflecting the planning culture more in its conceptual way (e.g. centralization vs. decentralization). Functional ties among the cities and regions, their division of labour and production as well as the nature of administrative systems are reflection of this layer in planning cultures.

Social and Behavioural Symbols

Social and behavioural codes are heavily influencing the planning culture. City or region is a social system, generating own symbols of social status, behavioural patterns, or other forms of symbolical expressions. Symbolical status of mayors, spatial planners, researchers, local authorities or municipality officials, their reputation and image, behavioural patterns, and codes of conduct deliver help to structure the social as well as spatial dimension. Managerial styles, decision-making, democratic tradition, or strong leadership are vigorous parts of planning culture and are observable expression of deep underlying assumptions and human values. Very important role in this process has been played by media.

Visual and Cartographic Representation

Integral part of each planning culture is the visual and cartographic representation of territory and territorial appearances. Maps are key elements in process of interpretation of planning vision, explaining the planning process and fostering the involvement of various actors. Communicative power of maps and visuals is extraordinary important especially in participative planning culture. Planning policy documents and visuals are not mere visual aids, but a powerful tool representing the existing planning culture within the territory.

Language

Language has turned to be key point of any social process. We can track the precursors of this tendency in the history: from the "turn to language" in the Vienna modernist movement in the 1920s till the "communication turn" of the European planning paradigm in the 1990s. Language is one of the most substantial frames of planning culture, offering the easiest and most perceptible form of symbols to encode/decode the content of planning culture and to transmit its narratives.

8.6 Essence of Leadership

Exact definition of the essential scope of leadership is almost impossible to find. Multitude of studies and researchers have tried to formulate definitions, but the approaches are dependent on the fact, which aspect of leadership stands in the spotlight, whether the leadership is perceived as a rather instrumental or rather substantial issue and many other aspects. The term "leadership" originates from old English word "lead" meaning "journey", "path", "road", so basic semiotic essence is based upon the idea of movement and guidance. Katz and Kahn (1978) defined leadership as "the process of influencing activities of an organized group towards goal achievement". Peters and Austin (1985) articulated leadership as a process "unleashing energy, building, freeing and growing". According to Rauch and Behling (1984), leadership is "the process of influencing the activities of an organized group towards goal achievement". One of the renowned gurus of research dedicated to mutual link between leadership and culture, E. Schein (1992) sees the essence of leadership in the ability "to step outside the culture and to start evolutionary change processes that are more adaptive". This approach of leadership is highly relevant especially in our field of research, namely the link between planning culture and leadership. Yukl (1989) defines the leadership as "the process whereby one individual influences other group members towards the attainment of defined group or organizational goals". All these definitions are outgoing from one assumption: individual variable (leadership) is set up into the group (culture, organization, profession) context.

The link between culture and leadership is obvious—culture generates extraordinary leaders and each leader is born and grown up in certain cultural framework. According to Schein (1992) leadership is never externalised in vacuum, it is always embedded in certain culture. This link is more transparent and easier to recognize in situations where certain culture is easy to distinguish against outer environment and where is present a strong inner identification with certain cultural patterns (e.g. corporate culture, club culture). In terms of planning, the situation is more unclear because of fuzzy and diffuse borders of certain cultural frameworks related to society, decision-making, and planning in general. Moreover, there is an immanent clash between leadership and planning culture—planning is still rather normative and regulatory, but leadership goes far beyond the prescriptive routine. The agility and strength to overcome the barriers of own planning culture is essential. Leadership obtained new dimension in participative/communicative planning cultures. Planning is not perceived as a purely professional expert activity but is approached as a collective effort coordinating interests and goals of multitude of actors. Highly profiled leadership, sensitive to particular needs of certain group of actors might be decisive breakthrough factors even in the dead-end and non-perspective endgames. In addition, the role of the leaders is not reserved for professionals; every single actor in planning process might step out as decisive leader.

Among the fields of planning culture, especially related to leadership, we can emphasize:

(a) *Degree of own initiative and pro-active behaviour of actors of spatial development.*
Open, democratic, and discursive planning cultures are based upon certain pro-active behaviour and leaders are those who set up these frameworks, challenges, and visions. Pro-active behaviour of leaders has highly motivating effect on the mobilization of entire planning process and makes people more eager, enthusiastic, and satisfied.

(b) *Risk taking behaviour and readiness to take action*
Leaders are waging the risk in situations when obvious routine is not enough to breakthrough. Readiness to action as a predominant trait of certain culture might mobilize all the actors and significantly increase "the pace" of planning.

(c) *Decision-making style and feedback mechanisms*
Democratic planning cultures are based upon the open decision-making style and ethical communication where the public is not the mere target of communication but takes responsibility also for adopted solutions. Leaders are those who are giving feedback, providing commitment and steering the planning process.

(d) *Integration and participation of actors*
Participation in planning process is never one single event but generates the background for long-term relationship. Repeated mutual communication and participation of various actors strengthen the inner social cohesion of the community and a society as a whole. Leaders are those who are searching for compromise, keeping the rights of minor actors and appeasing the hostile attitudes.

(e) *Degree of restriction and supervision*
Planning is always a voluntary, "controlled" process but the sense of autonomy is important for participation of various communities. Leadership might be key factor for mutual trust, delegation of various tasks as well for keeping the restriction in necessary limits.

(f) *Innovations and ability to learn and adapt*
Leadership is closely related to the ability to learn and adapt and to overcome anticipatory anxiety of failure. Concept of learning municipalities or smart cities is based upon the tolerance for errors and introducing new innovative approaches into planning practice.
(g) *Resilience of the culture*
Overall resilience of planning culture, its ability to withstand crisis situations and attacks from outside closely correlates with overall competence of its leaders. Leaders are those, who "defend" the planning culture in tough times, when underlying democratic values are doubted or compromised in any other way.

Opinions differ on whether the leadership is predominantly born or made. However, it is entirely accepted that good and mature leader must have the experience, knowledge, commitment, patience, negotiation skills, and especially trustworthiness (Amanchukwu et al. 2015). According to Gill (2011), the following leadership theories are the most influential in current leadership research and practice:

1. *Great man theories* assuming the essence of leadership is inherent—great leaders are born, not made. These theories do have its historical background and are portraying leader as an extraordinary heroic personality born to make the unique and non-repeatable decisions. Application of these theories in complex and multidimensional nature of planning culture is very limited.
2. *Modern trait theory* proposes that individual emerge as leaders across a variety of situations and tasks and utilize their personalities' potential in various contexts. Individual leadership traits include intelligence, adjustment, extroversion, openness, and general self-efficacy. Many people do possess traits typical for leaders, but they do not become leaders. In this group, we can include also behavioural theories suggesting that leadership requires a strong personality with well-developed positive ego and self-confidence.
3. *Contingency theory* assumes that different situations require for different characteristics of leaders—There is no single optimal profile for effective leader. These theories are better suitable for high complexity fields of action like spatial planning. Leadership is not the certain set of traits but kind of general ability to find balance between interest, various behaviour, needs, and context. Good leaders are those who are able to predict the future development in fuzzy and unclear situations, wage the optimal risk-taking behaviour or adjust their behaviour, and acting according to shifts of the current situation. Leaders are able to assess the needs of their followers and take stock of the situation. Success depends on a number of variables including the leadership style, qualities of the followers, and many unique situational aspects. Here we can include also the so-called situational theories, based upon the assumption that successful leader is able to "read" the situation.
4. *Participative theories* propose that successful leadership is based upon the ability of leader to involve and include the other into decision-making, bearing the responsibility and pro-activity behaviour. Leaders encourage participation and

contributions from group members and help group members feel more relevant and committed to the decision-making process. Participative/collaborative planning striving to include all the actors into the planning process is based on this type of leadership.

5. *Full-range theories* brought into the spotlight the term transformational leadership—Leaders are extraordinarily good in individualized consideration, intellectual stimulation, inspirational motivation, and idealized influence. In these theories, the process of leadership might be enriching on the individual level, satisfies the need for self-realization and commitment. These theories are better elaborated on the organizational level (leadership and corporate culture).

8.7 National Cultures and Planning

National culture is based on the belief of togetherness to ethnic group, civic group, or state. National identity is a bond binding the individuals to certain country or region. There are historical, geographical, cultural, political, and ethnic issues included. Historical experiences have shown that just countries and states sticking on to strategic visions and proven values are most successful in long-term period. Examples worth noticing might be an "American Dream", traditional and conservative values in UK or environmental values in Scandinavia. National cultures have largely influenced all cultural subsystems, including planning culture. Each planning culture includes contradictions between the past and future (Staniunas 2007). Centuries and decades had formed its current nature and shape accumulating and alleviating all the historical contradictions and conceptual collisions (Taššo 2008b). Even the role of the leader is heavily influenced by national culture framework.

Early research in terms of work with large samples has been focused on business cultures related to certain national cultural backgrounds. Hofstedé (see, e.g. Brooks 2003) work on impact of national culture onto business cultures was covering more than 15,000 questionnaires of IBM employees worldwide. His cultural framework, consisting of five dimensions (power distance, individualism vs. collectivism, uncertainty avoidance, masculinity vs. femininity, long-term vs. short-term orientation) became a benchmark for all other theories on impact of national culture onto various cultural subsystems. Large research of Shore and Cross (In Chin Mei Yen and Pulatov 2007) focused on the interlinkage between national culture and project group management, adopted six of his dimensions. Hall (1988, in Chin Mei Yen and Pulatov 2007) measured culture and communicative interactions based upon high and low context-related communication. Trompenaars (in Brooks 2003) introduced a survey on culture's influence on problem-solving. His approach is based on seven dimensions: universalisms vs. particularism, affective vs. neutral, individualism vs. communitarianism, achieved status vs. ascribed status, specific vs. diffuse, time as sequence vs. time as synchronization, internal control vs. external control. Further valuable research on national context of corporate cultures have been externalized by Denison (1990), Ouchi (1981), or Sännti (2001).

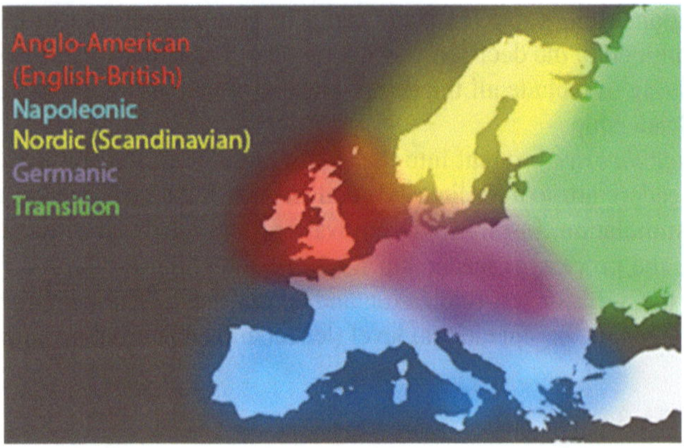

Fig. 8.1 Different institutional background for planning systems in Europe. *Source*: Authors, based upon Zweigert and Koetz (1998): An Introduction to Comparative Law, Third Edition, translated by Tony Weir

Planning culture is specific cultural subsystem because it is distinctive in its nature (compared to other cultural subsystems). While other cultural subsystems (culture of particular professions, corporate culture) are shaped predominantly by those living inside these subsystems, planning culture is semi-open system: it is shaped by the insiders (planners, architects, politics, lawyers) as well as by the broad public.

Historical development of European planning cultures and their backgrounds from the holistic point of view (see Figs. 8.1 and 8.2):

Planning culture has been advanced to be subject of several research activities, but the methodology and results of such research are quite scarce. Most studies rely rather on heuristic than empirical research; scientific methodology is lacking and outputs refer on single planning culture (Jaššo 2011). Friedmann (2005) requires "carefully designed comparative studies of how spatial planning is actually done (and with what results) across a spectrum of places, how it is understood by those involved in these processes and the legal and institutional foundations on which they are based". The following questions and open issues could be listed as the most important (according to Finka and Jaššo 2007):

- What were the predominant and underlying reasons of creating different planning cultures in Europe?
- What are the essential elements of difference?
- How our planning culture has been perceived by representatives of other distinctive planning cultures?
- Which elements of our planning culture are perceived as effective and which as non-effective (useless, obsolete, diffuse, incomprehensible)?
- Which elements of our planning culture have been adopted/adjusted by other cultures?
- Which elements of foreign planning cultures have been adopted by us?

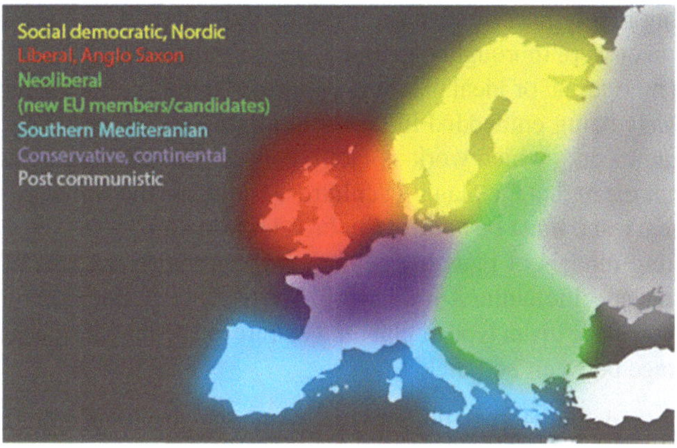

Fig. 8.2 Different background for planning systems in Europe given by diverse welfare systems. *Source:* Authors, based upon Duhr et al. (2010, European Spatial Planning)

- What is the role of the leader or are there any substantial individual roles within the process of developing planning culture?
- Which other planning cultures are compatible or having common heritage with our culture?

8.8 Our Model

Before designing any theoretical model of planning culture, we have to bear in mind the following underlying assumptions:

1. Old models were rather prescriptive and were often backed by old political clichés and narratives. For example, central European countries were often labelled to be "countries in transition" although this transition was concerning mainly the economic and social systems. Shifts in planning systems in these countries were rather evolutionary and smooth. In addition, old models and categorizations approached all former socialist countries as "one monolithic block", despite each one was displaying particular distinctions.
2. There is no clear-cut planning culture pattern in practice. Every culture involves different traits, characteristics, and hallmarks from various subtypes of culture, however one type dominates. Planning culture is predestined to involve internal contradictions (Staniunas 2007). Each particular planning culture might involve elements of high flexibility as well as low flexibility in different fields and dimensions (e.g. communication patterns, decision-making, risk taking behaviour). Even the democratic decision-making might significantly differ in its style and appearance (countries like Sweden or Switzerland differ in many ways, despite

both being highly participative and democratic). Cultural differences in project management were thoroughly analysed, e.g. in project CULTPLAN (2007).

3. There is no optimal or ideal planning culture. Planning culture is highly contextual, always being embedded in broader cultural, social, political, and institutional frameworks which work as facilitators or obstacle for any new element. It does not mean that the planning culture is static, but the shifts are more or less evolutionary and must be assessed in broader horizons.
4. No existing models of planning cultures are examining the role of the leaders within this specific cultural framework. We tried to outline specific leadership profile typical for given type of planning culture, respecting the necessary mutual interlinkage between both fields.

Our proposed model (see Fig. 8.3) is surveying planning culture within the sections of two main axes:

1. High flexibility vs. low flexibility
   (Flexibility reflects the ability of system to incorporate new impulses, provide feedback, and involve all the actors. It is not only the mere "pace of planning", but involves the openness of planning discourse and the ability to react).
2. Democratic vs. autocratic decision-making

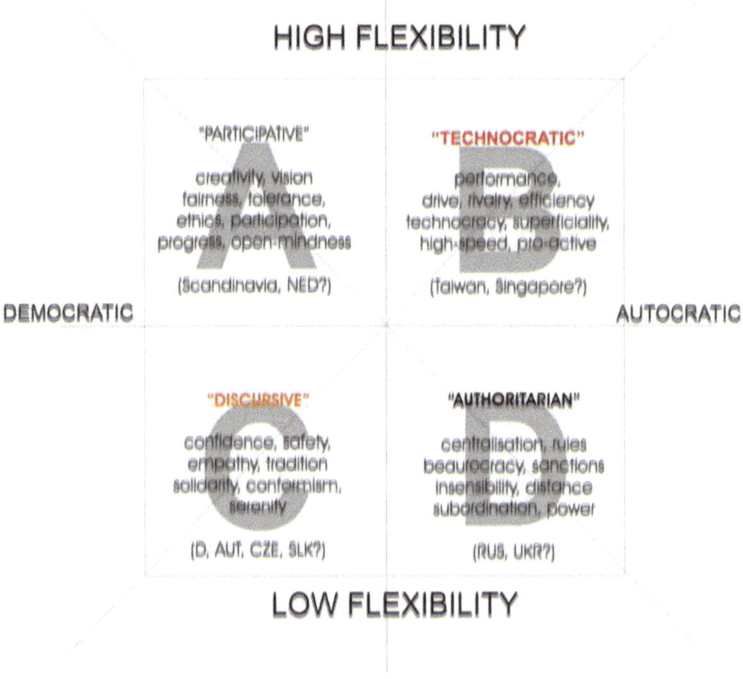

Fig. 8.3 The model of planning cultures. *Authors*: M. Jaššo, M. Finka

(This dimension measures indirectly the distribution of power, steepness of hierarchy within the society, and character of ideological background of planning culture implemented via followed value systems.)

The following prototypes of planning cultures emerge:

(a) Participative culture
Culture is focused on balanced and sustainable development, when the mutual consensus is the key milestone of effectiveness. Overall society is built upon the cohesion and balance and each change/shift is rather evolutional, taking into consideration the feedback and consensus.

Values:

Creativity, innovation, progress, shared vision, target orientation, fair competition, conception, tolerance, flexibility, ethics, beliefs, individual freedom.

Advantages:
Support of liberal and open thinking, proper conditions for creative development, agreeable social climate, honest and friendly social relationships, strong vision, new ideas and thoughts.

Disadvantages:
Lower efficiency of crisis management, too much convenience, weak motivation to withstand the conflicts, vulnerability of this culture under external pressure, slow decision-making, feebleness.

Successful leader is:

Independent, creative, visionary, open, team-oriented, supporting, productive, flexible, fair.

(b) Technocratic culture
Culture focused on results and progress; its success is measured by the economic efficiency and the small stakeholders do not have much to say. This culture is rather pragmatic and ideologically free. Sometimes, it is more "evidence-based planning" than "planning for the people".

Values:

Competition, rivalry, high efficiency, performance, endurance, individuality, drive, conflicts, motivation, pro-activity, superficiality.

Advantages:
Strong personalities could find the self-realization in this type of culture, support of individual and extraordinary performances enables excelling of prominent individuals.

Disadvantages:
Tensions, conflicts, interpersonal barriers, indifferent social climate, impersonality, and social distance.

Successful leader is:

Individualistic, calculating, tactic, closed, calm, inscrutable, professional, ambitious, and resistant.

(c) Discursive culture

Culture focused on integration and evolution. This culture has arisen by the modification of rather obsolete, normative, rational-predictive culture of modernity into more continental, communicative and sensitive one. However, transition was never completely smooth and this culture retained some "old features", mainly high degree of bureaucracy and high inertia.

Values:

Integrity, serenity, safety, confidence, cohesion, solidarity, conformism, lower flexibility, lower efficiency.

Advantages:

Social safety and social cohesion, backed by data and research, stability, efficient performance of routine tasks.

Disadvantages:

Lack of external impulses, lower self-reflection, stereotypes, routine, rigidity.

Successful leader is:

Responsible, considering, advisory, competent, empathic, coaching, rightful, awarding, loyal.

(d) Authoritarian culture

Culture is strongly related to national culture realms and the tradition of the current or former totalitarian regimes. Strictly bureaucratic culture based upon numerous restrictions, rules, and norms. Participation is rather low, and small stakeholders are outsiders in decision-making. Various power alliances are ruling the system.

Values:

Rules, centralization, bureaucracy, sanctions, insensibility, distance, subordination, regulations, authority, power, order.

Advantages:

Quick decision-making in crisis situations, unified external behaviour, the ability to keep the structure, resistance toward the external turbulences.

Disadvantages:

Lack of perspectives, stagnation, low motivation, alienation, pressure, lower self-realization, insensibility toward the small stakeholders, reign of rules and regulations.

Successful leader is:

Powerful, haughty, arrogant, restrained, formal, insensible, strict, blunt, authoritarian.

8.9 Central European Planning Culture

Analysis of previous spatial development in countries like Austria, Germany, Czech Republic, Hungary, Slovakia, or Slovenia has shown that there are many similar traits and characteristics appearing periodically in various situations and contexts. Not only history, but also planning systems and repertoire of planning practice mutually differ only in small peculiarities. They are being generated by similar systems of values, underlying assumptions and historical experience. Our debate on conceptual nature of the so-called Central European planning culture has been initiated in the seminar workshop Planning Culture in the Central and Eastern European Countries (29th June 2005) in Bratislava and has been continuously promoted within the various platforms (e.g. Spa-ce.net). Spatial development in above-mentioned countries has something similar, making them distinctive in comparison to other European states and regions (Finka 2006; Finka and Jaššo 2007).

According to the results from the research undertaken as stated in the introduction of this chapter, we may conclude that essential traits of Central European planning culture are clearly influenced by (according to Finka and Jaššo 2013):

- **Large time horizons of the planning executive**
  Central European planning culture has relatively "slow pace" and is rather adaptive, evolutional, and even bureaucratic. Planning process has many steps and time frames which are difficult to skip, accelerate, or modify. Rationalistic-predictive planning paradigm has never completely expired here, despite the progressive shifts in planning profession, giving more space and opportunities toward other, non-professional stakeholders. The planning bureaucracy has been always present at different levels of decision-making and executive. Planning always reflects the current ideological and political hegemony and division of roles in society. Central European planning was always more adaptive than trend-setting milieu.
- **Attention paid to social aspects of planning**
  Planning was always considered as a set of activities largely influencing social composure and coherence of society (planning was never purely "physical"), despite lack of coordination between land-use/urban/regional planning and other dimensions of spatial planning mix (strategical planning, human resource planning, landscape ecology planning). Lower social and professional mobility made the weakest stakeholders more vulnerable and exposed to negative consequences of neoliberal solutions. The overall situation after new millennium might be characterized by spatial polarization on different levels of society is weakening the overall (mainly social) cohesion. Social aspects of territorial and spatial processes are more visible than ever before (paradigm of new regionalism) and regional disparities are still growing from the 1990s and are one of the main scopes of the political and scientific discourse (Gajdoš and Pašiak 2006).
- **Growing general vulnerability of spatial/social systems**
  Acceleration of the spatial development has generated also the increased probability to face also the negative effects of the globalization as well as EU integra-

tion. Recent economic and financial crisis highlighted this risk and created threats which were not acute even some years ago. Fragile spatial and societal structures have been exposed to huge pressure originated either from international markets, unfavourable demographic prognosis, environmental hazards, or any other sources of risk. Social instability and deteriorating social cohesion became one of the most threatening factors. Sense of alienating and loss of the traditional social structure, loss of the regional identity, or dysfunctional social networks have hit some municipalities rather hard. At the beginning of the twenty-first century, the following risk-portfolio is essential (according to Gajdoš and Pašiak 2006):

- Risk of deepening of regional disparities, mainly in the field of the quality of human resources,
- Risk of the breakdown of social environment of marginalized regions,
- Risk of the breakdown of the readiness to implement regional policies (mainly due to the lack of financial and human capital).

• **Considerable methodical depth but lack of interdisciplinarity**
Planning "based more on data than on people" as the holder of knowledge brought precise classification, segmentation, evaluation, and passportization. Statistics, data, figures, and patterns were always the backbone of central European planning, giving little space for ad hoc solutions or creative adoption of new approaches. On the other hand, social sciences were involved rather modestly as if the "perfect planning was always normative planning in its inner nature".
• **Rather indifferent/sceptical attitude of public toward planner profession**

Public confidence, expectation, and trust are still lower than wished. Involvement of the public, various low-key stakeholders, and the presence of non-formal, more "role-based" planning models has been always rather scarce, even in Germany and Austria. In the former socialist countries (Czech Republic, Slovakia, Hungary, Poland), considerable distrust of public is the heritage of the late modernity under the socialism (1970–1980s) when alienation and the helplessness were the dominant emotions in public discourse. In these countries, even the new situation did not drastically change this attitude pattern—The decision-makers have been replaced but the fatalism remained (see Bútorová and Gyarfášová 2010). However, this is not typical solely for Central Europe, Staniunas (2007) writes, that planning culture in Baltic states is "culture of plans without planners" in which the plans got their own life and the planners were marginalized. This also indirectly indicates general lack of leadership, planning is depersonalized, de-emotionalized, and sometimes reduced to pure technocratic follow-up of routine and regulations.

Although the analyses have clearly shown the presence of dominant common features of the planning systems belonging to the Central European family, there are specific traits of the particular planning systems in particular countries which were highlighted for particular countries by the members of Central European Working Society of the ARL as follows (Roch et al. 2008):

- Overall planning culture and planning tradition in Hungary had been influenced by Christaller's model and Athens charter. Further substantial elements of the Hungarian planning culture are: (1) the so-called landscape thinking based on the landscape theories of Paul Teleki, (2) enorm importance of historically arisen spatial areas (megye, vármegye), (3) Hungarian functionalism, mainly influenced by concept of Zoltan Magyaris, (4) recent integration of main principles of EU regional policy (partnership, subsidiarity, programming), but in the practice there is still a lot of voluntarism (Dövenyi, in Roch et al. 2008).
- Czech planning tradition has been intensively shaped by the emphasizing of the role of the cities and their specific urbanity, protection of cultural and natural heritage, and respect to historical development. Among the most significant works depicting the Czech planning tradition might be the "pretty landscape" from J. K. Říha and "dwelling landscape" from L. Žák worth mentioning. First regional plans have appeared in 1950s and in late 1960s the concept of central settlements system. This system has been after 1989 political turnover abandoned and was replaced by rather neoliberal approaches (Rohrerová and Kyselka, in Roch et al. 2008).
- Regional development in Slovenia is based upon two principal concepts: (1) creation of polycentric urban systems and (2) the support of structurally weak and demographically challenged areas lacking behind in the overall development of the country. Cooperation among regional centres and small/medium-sized municipalities as well as the functional compatibility of various territorial subjects is essential. Planning system in Slovenia is based upon the similar values—The planning covering the whole territory of the state, binding character of the planning documentation, coordination of sector policies, horizontal integration, subsidiarity, participation, and polycentric development (Ravbar, in Roch et al. 2008).
- Planning culture and planning system in Slovakia have been largely based upon the common tradition of Austro-Hungarian empire (Theresian cadastre, land-use reform) as well as on the heritage of Czechoslovak functi,onalism. This heritage is highly influential in territorial division of Slovakia as well as in the various regional/urban planning documents. The central European culture leaves its track in the structure of planning documents, their content, planning process character as well as in the ex ante principle of planning. Planning was long years pure physical planning, without needed coherence and interlinkage to the different social, cultural and political processes running within the society. Current spatial/regional planning in Slovakia refers to values of sustainability and balanced development (Finka and Hrdina, in Roch et al. 2008).
- Spatial/regional planning in Austria has been externalized at various political levels, with the prominent roles of municipalities and regions (Bundesländer). Similarly to other central European countries, the main historical heritage has its roots in Austro-Hungarian empire (Theresian cadastre), though after world war

2, Austria went different way than the other adjacent central European countries. Central places concept of W. Christaller became an axis of spatial development in Austria after 1945. Today, the most influential values of planning in Austria are: Equal and socially justified conditions for every social group, competitive and innovative regions evaluating their inner potentials and territorial capital, sustainable, environmental friendly exploitation of natural resources, stabilization of ecological systems and smart, balanced use of spatial resources Austria (Schaffer and Beiglböck, in Roch et al. 2008).

8.10 Conclusions

Old models of planning culture were rather prescriptive and were often backed by old political clichés and narratives. For example, Central European countries were often labelled to be "countries in transition" although this transition was concerning mainly the economic and social systems. Shifts in planning systems in these countries were rather evolutionary and smooth. In addition, old models and categorizations approached all former socialist countries as "one monolithic block", despite each one was displaying particular distinctions.

There is no clear-cut planning culture pattern in practice. Every culture involves different traits, characteristics, and hallmarks from various subtypes of culture, however one type dominates. Planning culture is predestined to involve internal contradictions (Staniunas 2007). Each particular planning culture might involve elements of high flexibility as well as low flexibility in different fields and dimensions (e.g. communication patterns, decision-making, risk-taking behaviour). Even the democratic decision-making might significantly differ in its style and appearance (countries like Sweden or Switzerland differ in many ways, despite both being highly participative and democratic). Cultural differences in project management were thoroughly analysed e.g. in project CULTPLAN (2007).

There is no optimal or ideal planning culture. Planning culture is highly contextual, always being embedded in broader cultural, social, political and institutional frameworks which work as facilitators or obstacle for any new element. It does not mean that the planning culture is static, but the shifts are more or less evolutionary and must be assessed in broader horizons.

Planning culture is always arising on the interface between specific/unique versus universal. Planning as an integrated part of the societal reflection and society interactions with the environment generates unique reflections. Planning culture could be deemed as specific, unique as well as typical, non-casual set of approaches to the factors playing principal roles in the process of spatial development. It can be considered as one of the softest dimensions of spatial planning, being an important non-proprietary part of social and territorial capital. Current political development in Central and Eastern Europe seems to justify proposed model of planning and crucial role of political system with regard to multiple factors determination of planning cultures.

References

Amanchukwu RN, Stanley GJ, Nwachukwzu PO (2015) A review of leadership theories, principles and styles and their relevance to educational management. Management 5(1):6–14

Ash TG (1986) Does Central Europe exist? The New York Review of Books. Published 9 October 1986. http://www.nybooks.com/articles/archives/1986/oct/09/does-central-europe-exist/. Accessed 5 Dec 2014

Ash TG (1999) The puzzle of Central Europe. The New York Review of Books. Published 18 March 1999. http://www.nybooks.com/articles/archives/1999/mar/18/the-puzzle-of-central-europe/. Accessed 5 Dec 2014

Bělohradský V (2013) Mezi světy & mezisvěty. [Between worlds & interworlds]. Novela Bohemica, Praha

Brooks I (2003) Firemní kultura.[Corporate Culture]. Computer Press, Brno

Budaj J et al (1987) Bratislava nahlas. [Bratislava Loudly]. Slovenský zväz ochrancov prírody a krajiny v Bratislave, Bratislava

Bútorová Z, Gyarfášová O (2010) Slovensko na prahu tretej dekády slobody: kontúry participačného paradoxu. [Slovakia on the beginning of third decade of freedom: contour of participative paradox]. In: Plichtová J (ed) Občianstvo, participácia a deliberácia na Slovensku. [Citizenship, participation and deliberation in Slovakia]. Nakladateľstvo Veda, Bratislava, pp 137–184

Chin Mei Yen A, Pulatov B (2007) International differences in project planning and organizational project planning support in Sweden, Japan, Israel and Malaysia. Master Thesis, Umeå University, Umeå School of Business

CULTPLAN final report (2007) Cultural differences in European Cooperation. Learning from INTERREG practice

Denison DR (1990) Corporate culture and organizational effectiveness. Wiley, New York

Duhr S, Colomb C, Nadin V (2010) European spatial planning and territorial cooperation. Routledge, London

Finka M (2005) Vienna – Bratislava region: between cooperation and competition. In: Giffinger R (ed) Competition between cities in Central Europe: opportunities and risks of cooperation. ROAD, Bratislava, pp 58–65

Finka M (2006) Planning culture. Public lecture, IoeR Dresden 2006

Finka M, Jaššo M (2007) Planning culture as a factor of territorial capital. In: Spa-ce.net network conference Territorial Capital: guiding principle for European spatial development? Budapest, Hungary, 23 25 September 2007

Finka M, Jaššo M(2010) Selected aspects of territorial cohesion in Slovakia under recent crisis. Spatium international review. No. 23/2010. Belgrade, Serbia. pp 17–21

Finka M, Jaššo M (2013) Specifics and recent development of Central European planning culture. In: Spa-ce.net network conference. The role and future of spatial planning in Central, Eastern and South Eastern Europe, 25–27 September 2013, Dresden, Germany

Finka M, Kluvánková T (2015) Managing complexity of urban systems: a polycentric approach. Land Use Policy 42(2015):602–608

Friedmann J (2005) Globalization and the emerging culture of planning. Prog Plann 64(3):183–234

Fürst D (2007) Planungskultur. Auf dem Weg zu einem besseren Verständniss von Planungsprozessen?, PND Online III/2007, Available at www.planung-neu-denken.de. Accessed 6 Jan 2008

Gajdoš P, Pašiak J (2006) Regionálny rozvoj Slovenska z pohľadu priestorovej sociológie. [Regional Development of Slovakia from the point of view of Spatial Sociology]Sociologický ústav SAV Bratislava, Bratislava

Geertz C (1973) Interpretation of culture. Basic Books, New York

Gill R (2011) Theory and practice of leadership. SAGE Publications, London

Gottdiener M, Lagopoulos A (eds) (1986) The city and the sign: an introduction to urban semiotics. Columbia University Press, New York

Gullestrup H (2007) Intercultural understanding, culture theory and analytical models for empirical cultural studies. International Symposium "Planning Cultures in Europe", 28–29 June 2007, Institute for Urban, Regional and Environmental Planning, HafenCity Universität, Hamburg
Hansen KP (2003) Kultur und Kulturwissenschaft, 3rd edn. A. Francke Verlag, Tübingen
Hofstede G (2001) Cultures consequences, comparing values, behaviors, institutions and organizations across nations, 2nd edn. Sage Publications, Thousand Oaks
Hrdina V et al (2002) Joint chapter for the Vienna Bratislava region background report. Aurex, Bratislava
Jachna T (2004) Cyburban semiotics. First published at COSIGN, 14–16th September 2004, University of Split, Croatia
Jaššo M (2007) Competition and cooperation of the cities: Vienna-Bratislava Metropolitan Region. Proceedings from conference 2nd Central European Conference in Regional Science, Nový Smokovec, 10–13 October 2007, pp 354–365
Jaššo M (2008a) Cross-border co-operation challenges: positioning the Vienna-Bratislava region. In: Leibenath M, Korcelli-Olejniczak E, Knippschild R (eds) Cross-border governance and sustainable development. Mind the gaps. Springer, Berlin, pp 87–100
Jaššo M (2008b) Planning culture – soft tissue of Territorial Capital. Territorium No.8, June 2008, Institute of Spatial Planning at the Faculty of Geography, University of Belgrade, pp 43–56
Jaššo M (2008c) Plánovacia kultúra ako súčasť manažmentu priestorového rozvoja.[Planning culture as a part of management of spatial development] Forum Statisticum Slovacum. 3/2008. Slovenská štatistická a demografická spoločnosť, pp 77–82
Jaššo M (2011) Plánovacia kultúra. [Planning culture]. In: Finka M (ed) Priestorové plánovanie. [Spatial Planning]. ROAD, Bratislava
Katz D, Kahn RL (1978) The social psychology of organizations. John Wiley and Sons, New York
Keim KD, Jähnke P, Kühn M, Liebmann H (2002) Transformation der Planungskultur? Ein Untersuchungsansatz im Spiegel stadt- und regionalplanerischer Praxisbeispiele in Berlin-Brandenburg. Planungsrundschau – Zeitschrift für Planungstheorie und Planungspolitik 6:126–152
Knieling J (2007) Concept for an analytical model for research on planning cultures and first empirical findings of CULTPLAN. International symposium "Planning Cultures in Europe", 28–29 Jun 2007, Institute for Urban, Regional and Environmental Planning, HafenCity Universität, Hamburg
Knieling J, Othengrafen F (2009) Planning cultures in Europe. HafenCity University, Hamburg
Kuhn T (1970) The structure of scientific revolutions, 2nd edn. The University of Chicago Press, Chicago
Kühn M, Moss T (1998) Planungskultur und Nachhaltigkeit: Neue Steuerungs- und Planungsmodelle für eine nachhaltige Stadt- und Regionalentwicklung. Verlag für Wissenschaft und Forschung, Berlin
Kundera M (1984) Tragedy of Central Europe. The New York Review of Books, 26 April 1984
Mäntysalo R (2005) Approaches to participation in urban planning theories. In: Zetti, I, Brand, S (eds) Rehabilitation of suburban areas – Brozzi and Le Piagge Neighbouhoods, University of Florence, pp 23–38. http://virtuaali.tkk.fi/yhdyskuntasuunnittelu/ytk-eri/materia/mantysalo.pdf. Accessed 12 Feb 2008
Ouchi WG (1981) Theory Z. How American business can meet the Japanese challenge. Avon Books, New York
Peters T, Austin N (1985) A passion for excellence. The leadership difference. Warner Book Edition, New York
Pipkin JS, La Gory ME, Blau JR (eds) (1983) Remaking the city: social science perspectives on urban design. State University of New York Press, Albany
Raagmaa G (2002) Regional identity in regional development and planning. Eur Plann Stud 10(1):55–76
Rauch CF, Behling O (1984) Functionalism: basis for an alternate approach to the study of leadership. In: Hunt JG, Hosking DM, Schriesheim CA, Stewart R (eds) Leaders and managers:

international perspectives on managerial behavior and leadership. Pergamon Press, Elmsford, NY, pp 45–62
Roch I et al (2008) Central European planning culture. Position paper. ARL EAG Central and Eastern Europe, Hannover
Rode P(2006) City design – a new planning paradigm?. Discussion paper on urban age. London School of Economics. London http://www.urban-age.net/0_downloads/archive/Philipp_Rode-City_Design_A_New_Planning_Paradigm-DiscussionPaper.pdf. Accessed 12 Feb 2008
Sännti R (2001) How cultures interact in an international merger. Case MeritaNordbanken. Academic Dissertation, Acta Universitatis Tamperensis 819, University of Tampere
Sanyal B (ed) (2005) Comparative planning cultures, New York. Routledge
Schein E (1992) Organizational culture and leadership. Jossey Bass Publishers, San Francisco
Schimak G (2003) Leitbildartige Zielvorstelungen der Region Wien Bratislava. Expertinnenworkshop II. Hainburg, 25.11.2003. Projekt KOBRA, TU Wien. (Vision-oriented Goal Setting for Region Vienna Bratislava)
Staniunas E (2007) Spontaneous notes on planning culture of the Baltic States – case of Lithuania. International symposium "Planning Cultures in Europe", 28–29 June 2007, Institute for Urban, Regional and Environmental Planning, HafenCity Universität, Hamburg
Steinhauer C (2011) International knowledge transfer - analysis of planning cultures. Proceedings from REAL CORP 2011 change for stability. Lifecycles of cities and regions. http://www.corp.at/archive/CORP2011_87.pdf. Accessed 12 Feb 2014
Suitner J (2015) Imagineering cultural Vienna: on the semiotic regulation of Vienna's culture-led urban transformation. Transcript Verlag, Bielefeld
Umfrage zur Planungskultur (2007) PND Online IV/2007. http://www.planung-neu-denken.de. Accessed 6 Jan 2008
Vuolteenaho J, Ameel L, Newby A, Scott M (eds) (2012) Language, space, power: reflections on linguistic and spatial turns in urban research, collegium no 13 - studies across disciplines in the humanities and social sciences. Helsinki Collegium for Advanced Studies, Helsinki
Yukl G (1989) Managerial leadership: a review of theory and research. J Manage 15:251–289
Žigrai F (2005) Planning culture as a meta-scientific object. Seminar planning culture in the CEE countries. FA STU - SPECTRA Centre of Excellence, Slovakia, 29 June 2005
Zweigert K and Koetz H (1998) An introduction to comparative law, 3rd Editon, Oxford University Press, Oxford

Chapter 9
Urban Transformation - Cultural Heritage and Local Economic Development: The Public Sector on the Battlefront?

Robert Knippschild and Constanze Zöllter

9.1 Between Boom and Decline: Urban Development in Central, Eastern and South Eastern Europe

Capital cities in the European Union play a crucial role in the region's economic development: They are hubs for employment, sites of economic competitiveness, and drivers of innovation and growth. A comparison of the economic performance of cities in EU member states indicates increasing disparities among the urban areas. In some cases, capital cities develop into "magnets for growth", i.e. they display a monocentric pattern of urban development, attracting the lion's share of investments and resources (European Commission/Eurostat 2017). At the same time, population decline in cities ("shrinking cities") is also, increasingly, an international problem (Fol and Cunningham-Sabot 2010; Panagopoulos et al. 2015; Wolff and Wiechmann 2017) in Central, Eastern and South Eastern Europe.

Concentrated urban development in capital cities and other rapidly developing urban areas is associated with negative effects; similarly, population loss and economic decline harm less favorably located cities. Shrinking cities face diminishing urban functions such as public infrastructure, high-quality services and various aspects of urban living; cultural heritage is also degraded through falling population and vacant buildings. The end result is a lower level of attractiveness and competitiveness along with prospects of negative development. At the same time, booming cities and regions around them experience negative agglomeration externalities

R. Knippschild (✉)
Leibniz Institute of Ecological Urban and Regional Development & Technische Universität Dresden, Dresden, Germany
e-mail: r.knippschild@ioer.de

C. Zöllter
Leibniz Institute of Ecological Urban and Regional Development, Dresden, Saxony, Germany

© Springer International Publishing AG, part of Springer Nature 2019

M. Finka et al. (eds.), *The Role of Public Sector in Local Economic and Territorial Development*, EAI/Springer Innovations in Communication and Computing, https://doi.org/10.1007/978-3-319-93575-1_9

such as rising real estate prices and rents, traffic congestion, land use conflicts and poor environmental quality.

Current disparities in urban growth are associated with social polarization and economic imbalance. Therefore, sustainable urban and regional development entails innovative solutions to the problems associated with urban shrinkage and congestion. One crucial question in this conversation is whether smaller cities can benefit by taking on some of the population growth of the larger, congested cities, thereby contributing to a more balanced urban system.

Our assumption is that as digitalization and Internet commerce proceeds, small and medium-sized cities will be revitalized as more people work from home or from non-fixed locations. In any case, cultural heritage as soft locational factor will become more important. Decisions on where to settle will be increasingly influenced by the desirability of living environments and less by work considerations. The attraction of smaller cities might increase thanks to their less competitive housing and real estate markets. Certainly, this is not the case for all such cities; yet those with an historic urban fabric, good environmental quality, a robust social and cultural infrastructure and a lively civil society could enjoy a renaissance, even in peripheral locations.

In this chapter, we take a closer look at the German city of Görlitz, which is enjoying moderate economic growth after a long period of population decline (55,359 as of June 2016; https://www.goerlitz.de/uploads/02-Buerger-Dokumente/Statistiken/Statistisches_Jahrbuch_2016_komplett.pdf page 21), and where new solutions for urban transformation are currently being tested.

9.2 Urban Transformation: Between Cultural Heritage Preservation, Revitalization and Local Economic Development

Increasingly, policymakers and planners are forced to develop strategies to deal with urban shrinkage. Retirement migration and health tourism can figure into such strategies. Shrinking cities, which are usually viewed as less competitive, could gain some advantage by attracting senior citizens: "By converting vacant space and improving access to high-quality green space, shrinking cities can attract retirees in search of quality of life, who in turn might represent a crucial catalyst in urban renewal for shrinking cities."(Nefs et al. 2013: 1455) Although this approach has never been an explicit strategy in Görlitz, the city has been an attractive retirement location as far back as the so-called *Gründerzeit* period of the late nineteenth and early twentieth centuries. In the past the city proved popular due to its high-quality green spaces in the central part of the city, an attractive environment with many spa resorts and winter sport facilities and a distinctive scenery in the hinterland as well as a convenient rail connection to the capital, Berlin.

Kinahan and Ryberg-Webster see historic preservation in the United States primarily as the domain of urban professionals, closely tied to city planning and development. While preservation is promoted as a key driver of urban revitalization, there is a lack of empirical research addressing this intersection. They call for an expansive research agenda to investigate the role of preservation in revitalization and to rethink preservation policy in the twenty-first century (Ryberg-Webster and Kinahan 2014).

For many years the central issue in conservation was to resolve the conflict between preserving the character of existing historic towns and the upgrading of building stock. Recently, however, the concept of "heritage" has superseded that of conservation. Heritage is now marketed as a product according to the demands of the consumer, i.e. tourists. In Görlitz as elsewhere, this has resulted in the commercialization of heritage over the values of conservation. Today, a major objective in the management and planning of historic cities is to ensure a symbiosis between the needs of tourism and the care of the urban heritage. Here, Nasser argues for a sustainable approach to the management and planning of heritage sites which is based on a community and culture-led agenda (Nasser 2003).

As a preliminary conclusion, we assume revitalization of an historic built environment to be a complex and conflict-laden issue. In particular, we highlight the dearth of research that treats the intersection of preservation and urban development. A further assumption is that soft factors, such as social environment, identity or cultural environment, gain importance at a time of digital transformation. Among these, cultural heritage can be viewed as a particular asset of small and medium-sized cities in Central and Eastern Europe, which often face stagnation or population decline and yet have significant stocks of historic buildings. One such city is Görlitz, where currently a new instrument to foster urban revitalization and heritage preservation is being tested.

9.3 The Dilemma of Revitalization and Heritage Preservation: The Case of Görlitz

With around 55,000 inhabitants, Görlitz is a medium-sized city with an outstanding ensemble of historic urban structures from the Renaissance and Baroque periods as well as from the *Gründerzeit*. The historic building stock of 4000 listed heritage monuments is unique in Germany.

The city is located in the eastern corner of Saxony. After World War II, Görlitz was split into two when the German–Polish border was shifted to the west and, consequently, the eastern part of Görlitz became the autonomous Polish city of Zgorzelec. Today, there is a strong partnership between the two cities. Görlitz still suffers from its peripheral location; specifically, there is poor access to Germany's core metropolitan areas, and the hinterland on the Polish side is not fully developed. Furthermore, the city has been afflicted by severe demographic and socio-economic problems for more than 25 years. Following German Reunification, in 1990, and in particular in the wake of the currency reform, many businesses quickly went bankrupt or were shut down. This resulted in massive job losses that until today have not been fully restored. A secondary effect of job losses was a shrinking population, which fell by around 25% due to out-migration and a significant drop in the birthrate since reunification.

Population decline has been accompanied by a higher proportion of vacant buildings, including within the historic stock. Görlitz had already been afflicted by vacancies in the inner city since before 1989; then, German Democratic Republic authorities constructed prefabricated housing estates in the outskirts as a remedy for derelict and virtually uninhabitable housing in historic areas.

Around 75% of the historic building stock has been restored since the mid-1990s, encouraging an influx of inhabitants to the inner city. However, the vacancy rate remains high in the building stock that has not been modernized (see Fig. 9.1). Many such buildings have remained vacant for more than 25 years, resulting in severe damage that in some cases has weakened the structural stability (see Fig. 9.2). Although little of the historic building stock has been lost, there is urgent need for action in order to preserve Görlitz's unique cultural heritage.

Today, the city receives a high level of recognition for positive as well as negative factors. On the one hand, (Görlitz, 2015) is known for its historic centre ("the most beautiful city in Germany"), for the border location and its strong cooperation with the Polish partner city of Zgorzelec ("Europastadt/Europamiasto") as well as a location for many international film productions ("Görliwood"). On the other hand, it is still associated with the negative impacts of its peripheral location, namely out-migration, population decline, economic weakness and poor economic prospects.

However, the real estate and housing market in Görlitz is currently enjoying something of a boom, with growing demand for high-quality housing as well as retail outlets in central locations. This trend is certainly supported by the low inter-

Fig. 9.1 Vacant buildings in the historic centre of Görlitz (IOER-Media/Knippschild)

Fig. 9.2 Vacant plot after collapse of a listed building (IOER-Media/Knippschild)

est rates of recent years in Germany and throughout Europe, encouraging investors to look for land and property beyond the core metropolitan areas, where investment options are becoming rare. A trio of factors—population decline, economic distress and vacant buildings—constitute an urgent threat to a vast array of urban historic buildings, not only in Görlitz but also in many other declining cities (also known as *legacy*, *shrinking* or *post-industrial* cities). Here, preservation orders apply to a wide range of city neighborhoods where the private sector is willing to invest (Ryberg-Webster and Kinahan 2017).

Furthermore, families and young people are "rediscovering" medium-sized cities. Again, this trend is due to the negative impact of rapid urban expansion in metropolitan areas, impacts such as skyrocketing housing prices, increased pollution and traffic congestion. The beneficiaries are cities with attractive housing markets, strong social infrastructure and lively and unique urban lifestyles.

Clearly, these developments provide Görlitz with an opportunity to attract residents and to overcome the challenges of population decline and vacant buildings by revitalizing the historic building stock. After seeing its population decline for over three decades, the city has enjoyed modest population growth since reunification. Now local authorities must resolve the dilemma presented by the historic buildings: they present both opportunities and challenges vis-à-vis revitalization. The enormous number of 4000 "individual listed" monuments under the strictest protection orders is increasingly proving a hindrance to new investments in the inner city. The designation is significant because in 1990 many historic buildings, in particular those from the *Gründerzeit* period, were in a poor state of repair and threatened by demolition. In order to save them, most were quickly listed under heritage preservation law. Since time was short, the easiest way to protect them was to designate them as *individual* listed monuments, even though the listing of building *ensembles* would have been more appropriate. The latter type of listing is the strictest form of heritage preservation, ensuring that virtually no changes can be made, even if only a single element is the object of the heritage preservation order.

This specific situation increasingly constrains revitalization in Görlitz since, despite demand for renovated flats as well as retail outlets in the inner city, investors are reluctant to purchase listed historic properties. Investors, planners and architects often lack sufficient knowledge concerning the status of heritage preservation, resulting in their withdrawing any investment initiative.

9.4 Beyond Routine: The Urban Transformation Matrix in Görlitz

Against this background, the municipality of Görlitz has elaborated an approach called the *Urban Transformation Matrix* aimed at revitalizing the historic building stock protected under heritage preservation law. The first aim of the matrix is to preserve as far as possible the historic building stock and the urban fabric of the inner city, thereby to avoid urban "perforation", i.e. the dissolution of urban structures through demographic decline and urban sprawl, a phenomenon affecting other eastern German cities with a similar set of development conditions. This helps to preserve and enhance the attractiveness and competitiveness of the city by improving its soft locational factors. Here the objective is to ensure that listed historic buildings are neither demolished nor altered in a way that changes their basic character.

The second aim of the matrix is to exploit current trends in the housing and property market. The idea is to invite open discussion of structural alternations to buildings in order to foster revitalization and occupancy by well socially positioned inhabitants not only in the case of individual buildings but also in the wider neighborhood, which in turn can promote further revitalization and revaluation of the entire urban district.

The logic of the matrix is to take into account heritage aspects as well as the proposed post-refurbishment function of a building *before* launching the approval procedure for construction work. Here the innovative element is that the matrix considers the planned occupancy of the modernized building and its expected impact on the urban surroundings. It should be noted that the matrix does not affect the legal procedure for acquiring a building permit. Rather, the approach is to facilitate informal communication between the private sector (i.e. architects, planners, developers and investors) and the public sector (i.e. local authorities for urban development and the preservation of monuments) before conflicts arise in this complex field or, indeed, investors lose interest.

The Urban Transformation Matrix takes four aspects into consideration (with underlying factors indicated in brackets):

1. Building typology and year of construction;

2. Urban design evaluation (building condition, homogeneity in the neighborhood, landmark function or identity formation of the building);
3. Assessment of the planned project (architectural quality, building culture);
4. Forecast of urban impact (impact on urban development, function as an exemplary development).

The indicated factors will vary according to three spatial categories, derived from a pre-existing zoning system (see Fig. 9.3). The three zones indicate the degree of urban consolidation:

1. Consolidated area (dark grey): high proportion of renovated houses, low building vacancy rate, balanced social structure;
2. Area undergoing consolidation (medium grey);
3. Non-consolidated area (light grey): high demand for urban restructuring, high vacancy rates, social segregation.

In accordance with the matrix, intensive structural upgrading is permitted in buildings located in less consolidated areas. Specifically, the matrix generates 12 categories of potential structural intervention for listed buildings. These range from

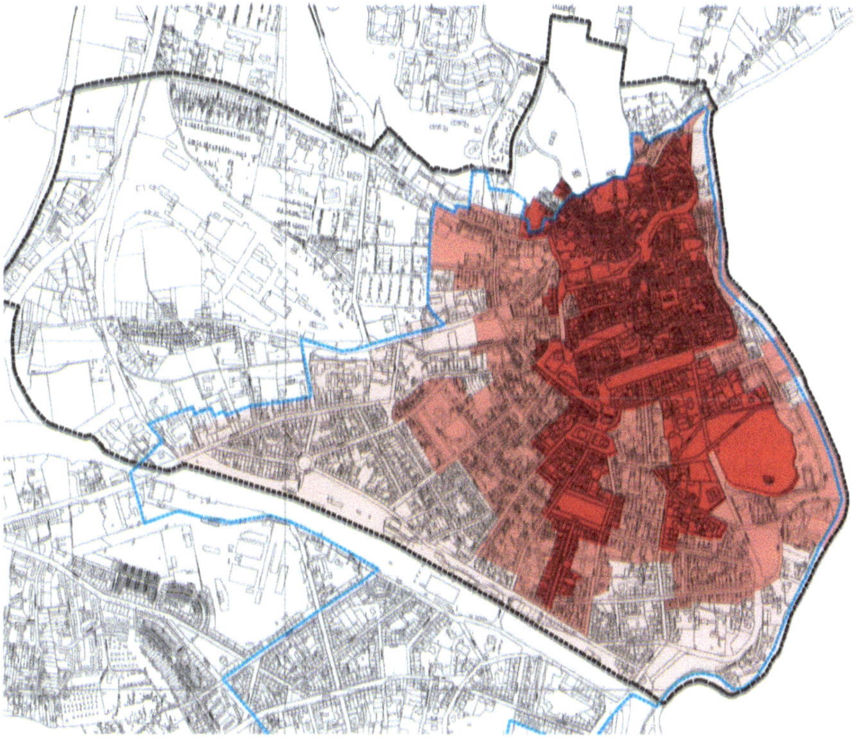

Fig. 9.3 Three zones of urban consolidation (Menzel 2016)

Fig. 9.4 The 12 categories of structural intervention (Menzel 2016)

"no intervention/change possible" to "demolition without replacement", with the aim of the matrix being long-term preservation rather than demolition of the historic building stock. Figure 9.4 presents visualizations of all of these resulting categories of intervention.

9.5 Difficult Application: Initial Reflections

From the beginning, the Urban Transformation Matrix was the subject of heated discussion by local residents and actors as well as by the public authorities responsible for heritage preservation. The conservation of Görlitz's urban heritage is a topic of public debate, reflected in numerous professional measures to protect the urban fabric and in voluntary activities and initiatives. Many of those involved in such initiatives were sceptical of the matrix, fearing an increase in demolition activities or massive interventions in listed buildings that would damage the city's unique historic centre. There was less heated talk within the municipal authority itself since public officials charged with heritage preservation are also responsible for urban development.

However, the agency at the state level responsible for heritage preservation feared a weakening of heritage preservation in Saxony. While understanding the unique case of Görlitz, its officials did not want the matrix approach to become a development model for other cities. After intensive hearings, the agency agreed to an experimental phase of 2 years to test the Urban Transformation Matrix and to evaluate the experiences made during this period. Subsequently, a council resolution was passed to approve the application of the matrix locally for a limited time period only.

Box 9.1 Approaches to Urban Development and Heritage Preservation in Germany

The development of cities with historic building stocks is the subject of various working groups in Germany, both at the national level and at the level of the *Länder* (federal states). The Working Group Historical Cities (*Arbeitsgemeinschaft Historische Städte*) was founded in 1973 as a cooperative venture between the cities of Bamberg, Lübeck and Regensburg to negotiate urban development that considers heritage preservation in historic inner areas. In 1991 this community was expanded to include the three Eastern German cities of Meissen, Görlitz and Stralsund. Until today the working group discusses sustainable concepts of urban development in established urban structures. At the level of the *Länder*, such working groups also exist in Brandenburg and North Rhine–Westphalia. Both groups have set themselves the goal of preserving and developing the cultural heritage of their cities. Within the community, the development goals are to be achieved through the exchange of experiences and information as well as by providing mutual advice and assistance.

In the building codes of the *Länder*, detailed regulations can be drawn up with the help of specific statutes for the preservation of design elements in buildings. These formal instruments are often combined with informal instruments such as urban design manuals (*Gestaltungsfibel*) or the use of board of advisers for architectural issues (*Gestaltungsbeiräte*). Such boards advise architects and investors as well as politicians and municipal administrators on issues of urban design. They merely make recommendations and are not authorized to make decisions. Currently, about 130 such boards exist in Germany as well as another eight temporary or mobile architectural advisory boards. Many of the boards can be found in North Rhine--Westphalia and Baden–Wurttemberg. In general, while architectural advisory boards are often based in growing settlements, shrinking communities also view this instrument as making a positive contribution to sustainable urban development (BBSR 2017: 25ff). The increasing number of advisory boards confirms that cities are ever more interested in preserving and exploiting their cultural heritage. As this is an informal instrument, there are significant differences in the structure and functioning of the architectural advisory boards.

Although this experimental phase has began in February 2017 and the project is scheduled to last until June 2018, some preliminary conclusions can already be drawn. The Urban Transformation Matrix is certainly an innovative instrument; in fact, a survey of approaches aimed at tackling the conflict between revitalization and heritage preservation in Germany has confirmed that no other model is comparable in form or complexity to matrix (see Box 9.1).

In general, the municipality of Görlitz is rather cautious about implementing the Urban Transformation Matrix. Indeed, the matrix is not well known among architects, planners or investors, and the city is doing little to raise awareness of it. Due to the limited resources of the municipal authority, a comprehensive application of the matrix is unrealistic. Therefore, the matrix can only be applied in selected cases. The aim of the matrix is to discuss possibilities of revitalizing listed but vacant buildings as early and as openly as possible while accepting certain structural interventions. To this end, dissemination of information among professionals and potential investors is crucial.

The Interdisciplinary Centre for Ecological and Revitalizing Urban Transformation in Görlitz, a joint research unit of the Leibniz Institute of Ecological Urban and Regional Development (IOER) and the Technische Universität Dresden, has helped organize a professional conference and public discussion on the topic of the matrix. In general, the reception among practitioners and scientists as well as the residents of Görlitz has been highly positive.

9.6 Conclusions and Need for Further Research

Since the test phase for approval of the Urban Transformation Matrix is not complete, any conclusions are still preliminary, and some questions emerged during testing which need further investigation. Yet, a few general conclusions concerning the matrix can already be drawn.

The matrix is an innovative approach to preserving cultural heritage while fostering urban regeneration and local economic development. Currently, we see no comparable approaches or tools in Germany. In the case discussed, heritage preservation and local urban development are not, per se, conflicting aims. Further investigation of the experiences gained in Görlitz is necessary to determine the general validity of the approach. A question of particular interest is whether the local government, which has the dual responsibility of heritage protection and local economic development, is able to play the role of facilitator between competing interest groups.

The matrix is a tool that specifically reflects the situation today in Görlitz, where the urban centre has a high proportion of vacant, mostly listed, historic buildings. As described above, action is urgently required in Görlitz due to the critical disrepair of many buildings as well as the current (and potentially temporary) interest of investors in smaller cities beyond the metropolitan hubs. Transferability of the matrix approach may, however, be limited due to the unique context of Görlitz. Further research is needed to determine how far and under what circumstances the matrix approach can be transferred to other contexts or extended to address additional aspects of urban development (e.g. energy efficiency).

Currently, we cannot say with any certainty whether the matrix is a generally applicable tool for assessing structural interventions in the building stock that accord with preservation orders. The same applies to the question of whether the matrix contributes to a reduction in the vacancy rate and thus to the revitalization and upgrading of urban districts.

Until now, local residents and the Saxon authorities have shown great concern that Görlitz's monuments be preserved. This concern along with a lack of support for the matrix approach at the local and state levels has fostered uncertainty within the municipality. Such uncertainty might be the reason the local authority has only cautiously promoted and communicated the advantages of the matrix to architects, planners, developers and investors within the city and beyond.

Development of urban areas should be done in a way that preserves the historical fabric of Görlitz. The promotion of this intention must be clearly communicated to the citizenry. The Urban Transformation Matrix is a communication tool that accompanies and fosters the formal process by which building permits are granted.

Finally, it must be pointed out that the matrix is not a product of academic or consultancy work but instead is elaborated by Görlitz's own Office for Urban Development, which is responsible for both heritage preservation and urban development. Although municipal staffing levels in shrinking cities are under pressure, innovation can be achieved in the public sector at the local level. The municipality of Görlitz now has to gain a certain level of self-confidence to better promote and communicate the matrix approach. Scientific support and monitoring can help disseminate the approach among academics and expert groups, but it does so without, unfortunately, reaching investors.

References

BBSR – Bundesinstitut für Bau-, Stadt- und Raumforschung im Bundesamt für Bauwesen und Raumordnung (BBR) (eds) (2017) Mehr Qualität durch Gestaltungsbeiräte. Perspektiven für die Baukultur in Städten und Gemeinden [Better quality through boards of advisers for design issues. Perspectives for building culture in cities and communities], Bonn

European Commission/Eurostat (2017) Urban Europe—statistics on cities, towns and suburbs—executive summary. http://ec.europa.eu/eurostat/documents/3217494/7596823/KS-01-16-691-EN-N.pdf. Accessed 11 Jan 2018

Fol S, Cunningham-Sabot E (2010) "Déclin urbain" et Shrinking Cities: une évaluation critique des approches de la décroissance urbaine. Ann Géograph 674(4):359–383

Goerlitz (2015) Goerlitz for many the most beautiful city on Germant. https://www.novyjicin.cz/wp-content/plugins/download-attachments/includes/download.php?id=29926. Accessed 31 July 2018

Menzel W (2016) Görlitzer Modell: Stadtumbau-Matrix. Bewertungssystem zur Weiterentwicklung innerstädtischer Bausubstanz [The Görlitz Model: Urban Transformation Matrix. An evaluation system for developing the inner-city building stock]. Unpublished presentation as draft resolution for the city council of the city of Görlitz. Accessed 4 May 2016

Nasser N (2003) Planning for urban heritage places: reconciling conservation, tourism, and sustainable development. J Plan Lit 17(4):467–479

Nefs M, Alves S, Zasada I, Haase D (2013) Shrinking cities as retirement cities? Opportunities for shrinking cities as green living environments for older individuals. Environ Plan A Econ Space 45(6):1455–1473

Panagopoulos T, Guimarães MH, Barreira AP (2015) Influences on citizens' policy preferences for shrinking cities: a case study of four Portuguese cities. Reg Stud Reg Sci 2(1):141–170

Ryberg-Webster S, Kinahan KL (2017) Historic preservation in declining city neighbourhoods: analysing rehabilitation tax credit investments in six US cities. Urban Stud 54(7):1673–1691

Ryberg-Webster S, Kinahan KL (2014) Historic preservation and urban revitalization in the twenty-first century. J Plan Lit 29(2):119–139

Wolff M, Wiechmann T (2017) Urban growth and decline: Europe's shrinking cities in a comparative perspective 1990–2010. Eur Urban Reg Stud 25(2):122–139. https://doi.org/10.1177/0969776417694680. Accessed 9 Feb 2018

Chapter 10
Social Innovation and Sustainable Economic Development: Participatory Tourism Destination Management

Aida Ciro, Merita Toska, and Peter Nientied

10.1 Introduction

The importance of tourism for the Balkan region is growing (World Travel & Tourism Council 2017). Tourist numbers are increasing, and, perhaps given the lack of much promising alternatives, governments embrace tourism as a priority economic development sector. There is however a substantial gap between governments aiming at tourism development, and tourism practices in real life. Tourism is a fragmented sector with many service providers and a dynamic tourism demand. Since tourists organize their travel increasingly by themselves with the support of internet (Nientied & Karafili, 2016) tourism demand also becomes more diverse and ad hoc. The increasingly important role of internet, the lower fares of airlines, and tourist searching for new destinations offer opportunities for the Western Balkan. However, Western Balkan governments have no adequate answers to such developments and for example marketing is rather conventional and destination management is inadequate.

A. Ciro (✉)
Department of Territorial Governance and Development, Permanent Workshop on Territorial Governance/Co-PLAN, Institute for Habitat Development/POLIS University, Tirana, Albania
e-mail: aida_ciro@co-plan.org

M. Toska
Department of Urban and Regional Governance and Development, Permanent Workshop on Territorial Governance/Co-PLAN, Institute for Habitat Development/POLIS University, Tirana, Albania
e-mail: merita_toska@co-plan.org

P. Nientied
Co-Plan Institute for Habitat Development, POLIS University, Tirana, Albania

Institute for Housing and Urban Development Studies, Erasmus University, Rotterdam, Netherlands

© Springer International Publishing AG, part of Springer Nature 2019

M. Finka et al. (eds.), *The Role of Public Sector in Local Economic and Territorial Development*, EAI/Springer Innovations in Communication and Computing, https://doi.org/10.1007/978-3-319-93575-1_10

The Albanian government recognizes the importance of tourism as an economic growth sector, yet, its contribution to the GDP is almost insignificant based on the National Institute of Statistics data. The latter may not be a genuine reflection of the real situation for numerous factors, among which informality. According to Medina and Schneider (2018) and Boka and Torluccio (2013), the informal sector in Albania accounts for at least 30% of the GDP. The World Tourism and Travel Council, however, assess that the direct contribution of the sector during 2016 is 7.9%, with a growth rate projection of 5.4% annually (WTTC 2016). Despite contradicting figures, the Government of Albania aims at developing tourism in its broadest sense and to extend the tourism season to span across all four seasons by introducing alternative forms of tourism, such as sightseeing, cultural, culinary, sportive, historical, religious, and adventurous. This would lift the development pressure off the coastal area and shift the focus towards more inland destinations.

The Government's overall development strategy is the National Draft-Strategy for Tourism 2014–2020. This strategy has two significant flaws. First, 3 years after discussion in parliament, it is still a draft policy. Insufficient attention has been given to the strategy after delivering the draft. In the real world, the importance of the policy is limited. Second, the strategy is broad and aims at promoting "everything". The strategy talks about sustainable tourism, but does not elaborate the concept in practical measures for destination management, environmental management, etc. In practice, it is noted that tourism development is rather uncoordinated, is led by the private sector tourism service providers, and the role of government at local and regional level is limited. A risk is that local governments (and national government too) are eager to get investors and can easily "forget" about sustainability when investors come with a tourism project proposal. Indeed, the current conditions for sustainable tourism in Albania are rather unfavourable. The current tourism policy of the government is inconclusive with regard to sustainable tourism development, the priorities for short-term economic development are high (and much higher than the concern for environment and communities), and local governments are weak in applying regulations. In terms of management and coordination, the culture of working together in partnerships in destinations is weak.

The experience of tourism in the Durres coastal region, close to Albania's capital Tirana, gives ample reasons for concern. Tourism development along the Durres coast is a typical example of unsustainable tourism as it is seasonal sun-sea-sand tourism, with high environmental costs and congestion. Local governments have issued permits for construction of hotels, and later "discovered" that their actions led to serious repercussions, such as higher investments needed in infrastructure. Alternatives for this type of tourism development are needed. There are however only very few best practices on small-scale sustainable tourism in Albania, and no replicable local sustainable tourism models have been designed and promoted. Exactly this issue is taken up in a sustainable tourism project developed in the Albanian municipality of Gramsh. The concept is to develop an innovation in local scale tourism, based on available resources and with local coordination. The role of an external NGO (non-government organization) is meant to fuel the process during the first 2 years and then develop others sustainable tourism cases. Before discussing the case of Gramsh, attention first turns to a brief discussion of the two key concepts of social innovation and sustainable tourism, and to the methodology of social network analysis, used for the present study.

10.2 Social Innovation and Sustainable Tourism

Innovation and Social Innovation

Innovation means the successful exploitation of new ideas, a process of turning ideas into reality and capturing value from them (Tidd and Bessant 2013; Tony et al. 2012). An idea, or "newness", is applied and implemented to create value. Without value creation, newness is limited to merely creativity. Innovation can be studied at different levels: is it new for a company, for a network, for the world? Newness of services, products, processes, and business models is in business management literature often defined at the organizational level. An idea may not be new to the world, but is new for an organization and its context (Nientied and Karafili 2016).

Traditionally, innovation has been seen as technological or product innovation, but correctly Conway and Steward (2009) distinguish a range of innovation types. Next to product innovation, they distinguish service innovation, process innovation, and organizational innovation, innovation of delivery, marketing innovation, business model innovation, and innovation of institutions. In this fashion, Hjalager (2010) identified five different types of tourism innovations: (1) product or service innovations; (2) process innovations; (3) managerial innovations; (4) marketing innovations; and (5) institutional innovations. The latter are very significant for tourism because innovation in tourism is quite different from innovation in other business sectors, due to the special features of tourism. The collective and fragmented nature of tourism and marketing, the people orientation, the intangibility, the easy imitability, and the global changes due to internet and lower costs of travel are among the features that render tourism and tourism innovation a special case. A firm-focused approach in tourism innovation would look at what, say, one tourism supplier in the business does. Tourism is a complex system of many small suppliers and other stakeholders like government, destination management offices, etc. and therefore the form of collaboration in networks and alliances is of critical importance. In this chapter, social innovation is considered to be a form of institutional innovation, new forms of collaborative and organizational structures that include social and territorial dimensions. This can be at larger scale, in regions, and small scale in communities.

Social innovation is a broad concept and has been critically discussed in the volume of Moulaert et al. (2013). We take it, rudimentarily, as a process of developing solutions to challenging social issues in support of social cohesion. Social cohesion is taken in a developmental way. We consider in this chapter the local domain, a municipality or similar territorial context, as the realm for social innovation. Phills et al. (2008) define social innovation as the process of inventing, securing support for, and implementing novel solutions to social needs and problems. Many more definitions are available, but in short, it comes down to social innovation as addressing problems, improving human conditions, setting agendas for a better future, and the like, as forms of social development.

Territorial Dimensions of Social Innovation

Social innovation entails new forms of collaboration and networking, with new dynamics of social relations. Social innovation is always in a social-cultural-political context that needs to be understood. And, social innovation is spatially embedded, in our case study this refers to the interactions and networks within the territory of Gramsh municipality and participation in external linkages. "The relation between social innovation and territory can therefore be defined as impacting on socio-spatial relationships in such a way that local autonomy is increased in the construction of futures" conclude Van Dyck and Van den Broeck (2013, p. 137). They consider local autonomy as capacities and more independence, rather than administrative autonomy, because autonomy without dynamic resources will not give much results. Such relationships, in addition to Costa and Brandão's (2011), suggested dimensions of territorial innovation models, build upon: augmentation of knowledge creation and spill-overs, growth-oriented capital, embeddedness of relationships, community mobilization, and involvement, which are shaped and driven by the enabling role of local institutional dynamics. As in the case of Gramsh, innovation resulted from the dynamic combination of the aforementioned dimensions with the local community serving as a fermenting agent for the socio-economic development of their territory.

Rodríguez (2009) states that socially innovative practices enable more integrated approaches to spatial development, by empowering local communities and encouraging them to participate in the decision-making process. In social innovation, an integrative approach is required for complex social issues. We will see later in this chapter that for local sustainable tourism development, the community is the starting point, but tourism development and planning also demands for new governance relations between local and regional tourism destinations and local stakeholders. This, in the case of Albania, implies national government structures, in view of the absence of regional agencies for tourism development.

This chapter blends social innovation with *sustainable tourism*, which in its turn is also far from unambiguous notion. Sustainable tourism has been defined in many ways, including elements of planning, environment, man-made heritage, ecology, social equity and participation, economic and social sustainability, and longer term future. Aall (2014) reviewed the concept of sustainable tourism and distinguished environment-sensitive tourism (reducing environmental impacts) and environment-dependent tourism, utilizing the environment as a resource basis for tourism. Here, we use the more general notion of UN World Tourism Organization Network for sustainable tourism: it takes full account of its current and future economic, social, and environmental impacts, addressing the needs of visitors, the industry, the environment, and host communities (cf. Beyer 2014; European Union 2006; UNWTO 2013; Nientied et al. 2018). A lack of precise definition and sustainability standards does not have to hamper the development of responsible tourism practices. Indeed, various authors in the recent volume edited by Singh (2017; cf. McKercher and Prideaux 2014) confirm that academic work on sustainable tourism has been highly theoretical, has been inward looking and has been telling the tourism world what is wrong and what it should do, rather than working on having an impact on the real tourism world. We adhere to this reflection and believe that developing sustainable tourism practices is more important than adding new details to the theoretical discussions. In other words, while the concepts of social innovation and sustainable tourism inform the present case study of tourism in Gramsh, our intention is practice-orientated rather than theory development.

Social Innovation for Participatory Tourism Development

The government in Albania is far from a convincing actor in the development of sustainable tourism. At local levels, capacities to manage territories as tourism destinations are limited. In 2015, a process of municipal reorganization took place, increasing the territories of municipalities (from 373 municipalities and communes consolidated into 61 municipalities), without enhancing local competencies. It is true that a decentralization process has been taking place in Albania for over a decade, but the financial and institutional resources of lower governments are still very inadequate. In this context, there is ample space for development initiatives from the people, if they are somehow guided. In other words, slow-to-respond formal institutional mechanisms pave a way for participatory approaches.

If the saying holds that "poverty leads to creativity", then the Gramsh case is a fertile ground. There are few development opportunities in the region, the disparities within the region are high, and there is very limited investment capital available. On the other hand, some limitations also provide opportunities: in the absence of a long-term national vision for development, a laissez-faire environment has developed, as well as a need-driven small-scale entrepreneurial spirit. Needed is an innovative approach to "put the dots together". Social innovation means in this territory a participatory approach to improve economic development and thus incomes. Costa and Brandão (2011) discuss public participation as social innovation and conclude that there is no standard model or a single approach for community participation in tourism development. Similarly, to any territory-related process, it will depend on local conditions, specificities, and embeddedness of relationships and businesses.

Public participation in tourism development is broader than participation in local government decision-making. Participation is coordinated local action in tourism development—participation is not just participation in decisions, but means undertaking bottom-up actions. There are many reasons which such participation should gain importance—and it does. Costa and Brandão (2011) highlight the economic potential and the improvement of human realization and social growth. One may also consider pragmatic factors stimulating participation in action: the conditions of weak or withdrawing public sectors, the context of internet facilitating access to the tourism market and increasing fragmentation, and opportunities for tourists to enjoy authentic experiences in a sustainable environment, which they get less in more formally organized and larger scale tourism.

10.3 The Methodological Approach

The approach used in the case of Gramsh local tourism development, consisted of two main elements: building social networks and participation to build trust. The social network approach employed in the case study of Gramsh combines elements from the social network theory by Pan and Crotts (2012) with the Kline and Rosenberg (1986) chain linked model. Based on Pan and Crotts (2012), social network theory views the community of individual actors as nodes. Each node is connected and interconnected to other nodes at different levels, contributing to the shaping of an open social network. Within a social innovation and sustainable development context, this social network could comprise of nodes such as local community, private sector, local authorities, media, civil society organizations, and academia. The establishment of relationships between these nodes is what creates "the open social network", and their alignment towards a shared vision/aim is what creates value for a specific process, i.e. sustainable tourism development.

The relationships and processes within the open social network are not linear, but rather complex, uncertain, somewhat disorderly, and subject to changes of many sorts as Kline and Rosenberg (1986) note. In other words, if the social network theory helps identify the key actors in a process of social innovation within the sustainable tourism development discourse, the Chain-linked model by Kline and Rosenberg (1986) help examine the information/cooperation exchange on an intra and inter-network level since innovation is not an isolated occurrence.

Instrumental to such an exchange is trust. Albania is a typical low-trust society, and some explanation of this phenomenon has been given by Nientied and Karafili (2016). In a small community with a rural culture and a backward economy, the trust of the population in institutions, in the community, and in others outside the own skin group is low. Yet, trust is needed for innovative practices because social innovation is essentially community based. Gramsh community members had low trust in the government, low trust in the NGO planning to help Gramsh, and low trust in each other in terms of collaboration. The NGO was quite aware that helping the community to build trust on basis of experience—not based on visions and nice words—was required. The approach was basic, matching the nature of trust as a basic feeling and attitude—as we will see in the case study when elaborating the action of "*the Gramsh breakfast*".

10.4 The Case Study of Gramsh

The Context

The Municipality of Gramsh is located in the central part of Albania and shares borders with seven municipalities, of which four constitute relatively important urban centres. Although parts of the city date back to almost 600 years ago, it was not until 1960 that Gramsh earned its "town" status, mainly owing to the advent of an industrial sector planned under the previous communist regime. Various experts and professionals from different parts of Albania were embedded in the newly created city to support the new town development process. The industrial activity constituted the key economic activity for the Municipality up until the 1990s, consisting of the *Mechanical Factory of Çekin* and the *Battery Manufacturing Factory*. These two establishments contributed to the consolidation of Gramsh as an industrial town. Both industries collapsed after 1992, leaving the city with idle plants and a negative image of a dirty post-industrial small town.

Following the change of the political system in 1992, and the territorial administrative reform (2014), which resulted in consolidation of the administrative units in the country, the Municipality of Gramsh was subjected to significant territorial changes: its territorial scope grew by 300-fold, from 2.48 to 737 km^2, its administrative composition expanded to include 10 administrative units, while population increased by 300%, to 36,184 inhabitants. The significant increase in territorial coverage and comparatively low population defines a relatively low population density at app. Forty-nine inhabitant/km^2, a challenge in terms of service provision. In the wake of such developments, the municipality experienced a shrinking economic activity with the collapse and dismantling of its former industrial activity, resulting in high unemployment rates particularly among youth with figures as high as 82.9%[1]and high migration and immigration. The latter has contributed to the depletion of active labour force, which accounts for a total 68%[2] of the population of Gramsh. In short, the challenges the new municipality is faced with are: bigger territories to manage, bigger population (low density) to serve, limited municipal budget, low valorization of its local resources, poor internal connecting infrastructure, etc.

On a positive note however, the Territorial Administrative Reform (TAR) contributed to the altering of the municipal profile, adding to the Municipality a considerable stock of natural and cultural monuments amounting to a total of 21 including natural monuments and geo-monuments. The Natural Park of Tomorr, the Sotira Waterfall, the Holta Canyon, Dushku Lake, the Cave of Kabash, Fortress of Tervol, etc., in addition to the overall attractiveness of the landscape, have considerably transformed the profile of the municipality, and subsequently its economic development profile, transiting from an economy based mainly on industrial activity and agriculture, towards agricultural activity and, very slowly, tourism development. However, the Municipality is lacking in experience and tradition in tourism development and management, and faced the challenge of managing tourism resources, particularly in terms of achieving a profitable utilization of its resources, yet ensuring preservation or use without exploitation. Angelevska-Najdeska (2012, p. 73) definition of sustainable development as "a balanced economic, social and cultural development without endangering the environment in which future generations will enable them to develop the same or higher than the current level", captures in essence the Municipality's desired approach to promoting Gramsh as a tourist destination.

[1] Local Development Operational Plan of the Municipality of Gramsh, 2015.

[2] Active labour force to be verified. It refers to the age-group 15–64 years.

Such a local ambition, willingness, and commitment to tourism development comes at a time when the Government of Albania has declared tourism a national priority and has initiated a number of processes to complete the legislative framework, such is the drafting and passing of the new law on tourism. An agreed strategic vision that aligns and guides all national efforts and at the same time streamlines national priorities with local ones, is still unclear, as the only Strategic Document on Tourism Development is still at a draft-stage, although it was designed to guide development in the sector for 2014–2020. Further, the operationalization of the draft Strategic Document, although a positive development in setting the framework for the sector development and management, does not suffice to ignite tourism development locally. It requires the presence of numerous enabling conditions such as qualified human resources in municipalities, intra-regional policy and local economic development coordination, involvement of numerous actors including government and donor assistance and private sector, as far as fulfilling of municipal functions in this regard, such as offering the necessary infrastructure for tourism development.

Amidst such developments and the challenge to orient local economic development in the rapidly expanded Municipality of Gramsh, in 2015 the Mayor expressed an interest to collaborate with POLIS University and Co-PLAN, Institute for Habitat Development. Both institutions are known for their technical expertise in strategic planning, economic development, and territorial governance. They were approached to assess and better plan the municipality's local economic development priorities. The focal point of interest for the Mayor was the planning and implementation of local economic development catalytic projects and initiatives. With a modest annual municipal budget of app. EUR 5.5 million in 2016, the formerly industrial Municipality of Gramsh was trying to explore and diversify its economic activities in view of its new territory-based resources, and the central government priorities. In more specific terms, the Mayor of Gramsh was particularly interested in:

(a) The drafting of a local development operational plan, as a mid-term (2–3 years) document to orient the development priorities of the municipality in the form of priority projects. This document would enable the municipality transit towards the preparation and approval of the General Local Territorial Plan (GLTP), both a binding document and a legal requirement, based on the Law no. 107/2014 "On Territorial Planning and Development".
(b) Promoting economic activities with a focus on tourism, in an attempt to reverse the high unemployment rates in the municipality, yet tapping into own natural and historical resources.
(c) Articulating and formalizing the municipal tourism ambitions in a Municipal Tourism Development Strategy.
(d) Diversifying and accessing alternative financial resources from third parties, including donor and development agencies to finance priority projects mainly in infrastructure, since current municipal budget is limited to covering operational and personnel costs.

The Municipality of Gramsh, or the region, did not have a good reputation or a brand, and its experience in tourism was limited to ad hoc visits organized by local tour operators to a waterfall situated in Sotirë, one of the ten comprising administrative units of the new municipality. Gramsh was not perceived as a tourist destination, and its former heavy industrial activity and poor infrastructure and connectivity, had contributed to a widely held perception of Gramsh as remote, polluted, unattractive, although the majority of people that were asked (random), had never actually visited the place. The perception was further reinforced by the negative environmental impact from the construction of a major hydro-power plant in the municipality, Banja Hydropower Plan, which diverted the natural river flow (Devolli River) resulting in an artificial lake (Banja Lake). Despite the environmental impact, the lake contributed to the landscape reshaping of the area, adding a water-front element to the city. In an attempt to counter-balance in part the negative environmental effects, the company agreed to invest on the improvement of the secondary inter-urban road connecting Gramsh to all other surrounding urban centres.

Through a collaborative process and a visioning exercise among the Municipality, Polis University and Co-PLAN, it became apparent that a shift was required, not only in the way the Municipality would choose to orient its economic development agenda, but also in the way it would promote Gramsh as a tourist destination, and the type of tourism activity it would choose to develop.

From the outset, it was jointly agreed by the three parties involved, that any strategic documents and processes undertaken would avoid the standard-formal public sector top-down approach, with most documents ending up unused. The design and preparation of both the "Local Operational Development Plan" and the "Municipal Tourism Development Strategy" would employ a strong participatory element. The community would have to feature an important part in planning and decision-making rendering the process complex given the novelty of such a practice in Albania, and the low level of confidence among them in such initiatives—later also confirmed by the survey results.

The collaboration model is based on the specific complementary contribution, skills, and experience that each actor can bring to enable the accomplishment/realization of the process/objective. In the case of the Municipality of Gramsh ambition to assess and better plan its local economic development priorities, the constituent parts of the collaboration were:

1. The Municipality of Gramsh—the public entity in the capacity of the assistance requesting party, required by law to ensure the administration and management of the administrative unit.
2. POLIS University—the academic institution ensuring knowledge transfer and state-of-the-art research, ultimately aimed at the improvement of the public policies. In this case, the university would contribute to technical knowledge from the realm of spatial planning.
3. Co-PLAN, Institute for Habitat Development—a non-governmental organization facilitating processes and transfer of applied knowledge capitalizing on long-term experience in community development. As a constituent part of the model, Co-PLAN's input concerns dynamic participatory processes, strategic planning, and local economic development.

This partnership all in all served to streamline the local economic development assessment and planning process. It combined clearly distinct skills and expertise of each constituent part of the model, resulting in processes and outputs that build on a genuine reflection of the context.

All considering, the collaboration adopted a threefold approach:

1. Desk-review of all documents including planning, operational, strategic, medium-term budget programmes of the municipality and of the constituent administrative units to date. Through this process, it was possible to assess whether these documents had been operationalized (in part or in full), the extent to which they had contributed to addressing the then-current needs and priorities, and what gaps/issues had resulted in the process.
2. Two rounds of surveys were conducted:
   (a) Using a convenience sampling method, a survey was aimed at the local business community to assess the supply side in terms of service provision capability and capacity, growth potential, diversification of current offer, willingness to become part of a local value chain and contribute to sustainable local economic development, ability to interact and network with external actors, ability to promote its services and reach-out prospective customers, the extent to which they view Gramsh as a tourist destination, and they feel connected to the territory.
   (b) A second, simple visual-based survey aimed at assessing the demand side. Random people aimed at assessing the perceptions of Gramsh as a tourist destination. Three questions were asked, namely: (1) Would you go to Gramsh? (2) Why (not)? (3) Would you go to this place showing on the picture (respondents see a picture with Gramsh natural landscape)?
3. Site-visits in all the administrative units of the municipality aiming to explore and assess: (1) the territory configuration and functioning; (2) the territorial typologies; (3) existing stock of natural resources, location, and current level of valorization; (4) accessibility, connectivity, and quality of infrastructure; (5) proximity of service providers to natural/historical resources; (6) existing itineraries and potential for future growth; (7) the typology of touristic activities most suited to the territorial profile of the Municipality of Gramsh.

Parts of the process were documented and shared on social media in the form of a dynamic promotional campaign under the #EjaNëGramsh denominator aimed at: reversing the poor perception of Gramsh as a *dirty post-industrial town into* a *tourist destination*, while simultaneously introducing it to the potential domestic demand. The #EjaNëGramsh brand was designed and used as a multipurpose instrument: (1) provide support to the community mobilization process by increasing local pride and self-identification; (2) engage domestic demand particularly among tour operators and domestic tourists; (3) to document the initiative in a way that is traceable, replicable, and very importantly open to all skeptics to view the progress of the process incrementally. The hashtag became so popular that visitors would independently use it when posting or sharing images or posts related to Gramsh.

Parallelly, in order to trigger the first batch of domestic tourism demand, yet at the same time test the local service provision capabilities, a recurring hallmark event was conceptualized, "Natyra Fest". It would bring together elements from both local culture and active tourism alternatives giving rise to the first Gramsh-based touristic offer. These kind of hallmark events are deemed critical to putting a newly emerging tourist destination on the map. When large enough and prominent enough, these events grow into attractions in and of themselves, and drive tourist traffic to the city and outskirts, including local businesses and attractions. A number of studies have shown that hallmark events generate high tourist demand as well as high value (Getz 2008). In fact, some examples are so successful that they become associated with the city and thus require only minimal advertising cost once the association is established. Easily recognizable examples include: the Lysefjorden Hiking Festival[3] in Norway, the Glastonbury Festival in the United Kingdom,[4] and the Apollonia Art Festival[5] in Sozopol, Bulgaria. In the case of Gramsh, the key features of the proposed hallmark event "Natyra Fest" were:

- Selected offer of local culture and heritage elements with a wide enough appeal for general audiences, attractive for both domestic and foreign tourist;
- A multiple day duration, such as a long weekend, to encourage over-night stays by tourists
- Strategically planned to formalize and extend the tourist season and offer
- Open to local tour operators, producers, artisans from the neighbouring regions
- Stimulating linkages to other economic activities such as food and beverage, camping and local transport, and specially designed experience such as apiculture

In addition to the dedicated brand #EjaNëGramsh, an intensive communication campaign was undertaken with the public, employing online and traditional mainstream media. The campaign aimed at initiating a discussion on Gramsh as a tourist destination specifically targeting those who had never heard of Gramsh before, and those who had a negative perception of Gramsh because of its industrial past.

The Application of the Three-Partite Model to #EjaNëGramsh

Based on the preliminary cognitive process, it was possible to collect all the necessary data that would help in shaping and applying the proposed model. Some of the key findings that contributed the most were:

1. The lack of an approved General Local Territorial Plan (GLTP) in Gramsh had allowed for chaotic developments, in absence of a shared vision and strategy. Any documents found (not applicable, outdated) had no relevance for the existing

[3] https://www.hikingfestival.com

[4] https://www.glastonburyfestivals.co.uk/

[5] https://bulgariatravel.org/en/object/85/festival_apoloniya

context and municipal ambitions, given the significant territorial and administrative changes of 2015. The drafting and approval of a GLTP as well as a strategy to orient the local economic development was therefore an imperative.
2. Key findings from both surveys undertaken, identified:
   (a) Significant lack of trust and collaboration among local actors, who did not associate Gramsh to tourism and therefore saw no development potential. The levels of trust in any upcoming plans were so low that the response rate to the survey was at 10%.
   (b) The majority of people surveyed had no prior information on Gramsh, finding it difficult to associate Gramsh to a representative image. This finding was particularly helpful in conceptualizing the branding (#EjaNëGramsh) and communication strategy (hallmark event and media coverage) of Gramsh.
3. No connected dots: despite the plentiful existing stock of natural resources, the diversity in typology, the dramatic rich landscape, the yet-to-be-explored itineraries and its good connectivity with other surrounding urban centres, there were no value chains or any attempts to valorize any of the aforementioned advantages in function of local economic development.

The model, similarly to the Kline and Rosenberg's (1986) concept of open social network, focused on building relationships and processes, not linear, but rather complex, uncertain, somewhat disorderly, and subject to changes of many sorts. The model implied that work would have to be carried out simultaneously with various actors (nodes) and engaging in various processes and relationship building (lines). The model applied in Gramsh consisted of six key steps (see Fig. 10.1): (1) Identification of the Actors; (2) Dialogue initiation; (3) Establishing relationships; (4) Agreeing on a commonly shared vision; (5) Orienting the economic activity towards a shared vision (map of the vision); (6) Improving and expanding the existing tourist supply.

Identification of the Actors. Instrumental to a participatory process, the step of identification of actors ensured to scan and map all possible actors (nodes), and their respective networks (entirety of relations established with their respective collaborators) who could jointly influence or would be influenced by the upcoming development processes (see Fig. 10.2). Some of the key actors identified were:

- Local businesses (the nodes), essential in terms of products and service provision particularly in the fields of hospitality, local transport, and tour-guiding and camping equipment. Special place-specific activities which would enrich a tourist's experience, such as apiculture, artisan bakery, and diary production, as well as home-distilled spirits and wines were also mapped. Following the identification of the *"nodes", respective* networks were also mapped. Such networks included supporting services to the local business community: accounting services, marketing, printing and publishing, suppliers of raw and packaging materials, etc.
- Local community (the node), the subject of any upcoming development, and its twofold role in both designing the destination's future, as well as contributing to the creation of a suitable climate for hosting visitors. This particular actor group

Fig. 10.1 Chain-linked model as applied in #EjaNëGramsh (Source: the authors)

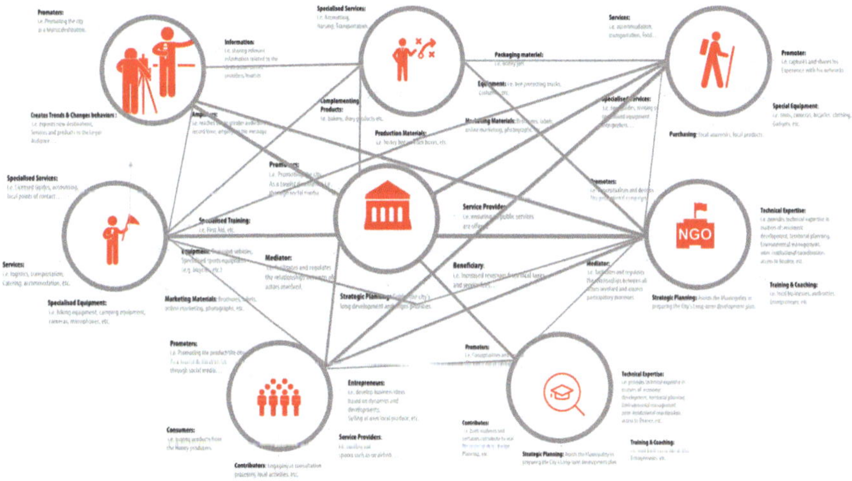

Fig. 10.2 The open network system in the case of #EjaNëGramsh (source: the authors)

was mapped in its capacity as: *contributors* to consultation processes and local activities; *promoters,* of the city touristic offer through various channels including social media interaction; *service-providers*, renting out spaces through various platforms, or artisan work; *potential entrepreneurs,* developing business ideas based on the dynamics and opportunities.

– Local authorities (the node), the authority mandated to guide the city's development and ensure alignment of immediate priorities with long-term vision; in the role of a mediator facilitating and regulating the relationships between all parties involved; local public services provider and promoter of the city as a tourist destination; beneficiary of the potentially increased revenues from local taxes and service fees.

- Tourists, the node setting all actors in motion, by shaping tourism demand, more specifically in terms of: purchasing of local products and services—accommodation, transportation, food and beverages, including specialized services and equipment; also mapped for his promoter role, who in the process, captures and shares his experience with his network.
- Tour operators dealing especially with hiking and other outdoor-based activities, a crucial node in shaping the touristic supply by integrating Gramsh in their touristic offer; by promoting existing itineraries and experiences that are place specific, and exploring new ones; contributing to the shaping of a value chain by building relations with local businesses along the itineraries, or promoting entrepreneurial activities in areas that are poorly covered with services. This node is particularly important in the case of Gramsh because of: (a) limited knowledge and information on Gramsh as a tourist destination; (b) the outdoor profile of Gramsh' touristic offer.
- Technical NGO (both a node and an actor) providing technical expertise in matters of: local economic development, territorial planning, environmental management, inter-institutional coordination, access to finance, etc.; facilitating and mediating inter-actor relationships and assuring participatory processes; training and coaching of both local authorities and local business communities; promoters of Gramsh as a tourist destination, conceptualizing and designing the overall communication campaign, including #EjaNëGramsh and Natyra Fest' hallmark event.
- University (both a node and an actor), assists the municipality with technical expertise in preparing the city local operational development plan; contributor with both students and lecturing staff to processes such as design, planning, and promotion.
- Media including traditional mainstream and social online-based media, a node instrumental in exposing Gramsh as a new upcoming tourist destination; in amplifying this message by reaching out to greater audiences in record time; in sharing relevant information related to the destination, service providers, and tourists. Social media particularly important to Gramsh-based service providers as a low-cost-high-reach promotional tool for their products and services, enhancing both visibility and business credibility.

Dialogue initiation among the actors. All the research work conducted to better understand the territory, resources, actors, and potential for development, spoke of all elements necessary for the creation of a unique tourist experience being there, yet, not streamlined, valorized, or organized into a value chain. Consequently, the model's second step focused on initiating dialogue among the various actors identified, ultimately aimed at: increasing trust, exploring and establishing collaboration avenues, encouraging them to be an active part of the planning process. Fundamental to any participatory process, this step was used to convey to the actors the importance of their input into the evolving stages of the process, as well as to showcase an application of participatory planning in real life. Such an approach gave sufficient time to the sceptics to better understand and consider an eventual involvement in this process. The municipality's initiative to promote Gramsh as a tourist destination in an effort to trigger local economic development was publicly introduced in December 2016 through an open invite to the local community and interested actors to become an active part.

Establishing relationships. Once the initiative was introduced to the actors, they were invited to reflect on the most suitable forms of development, which were inevitably conditioned by their interests, capacities, and profiles. The initiative although initially met with surprise, and some degree of mistrust, was well received considering that, up until that point that Gramsh as a tourist destination was simply a theoretical discussion. Considering the positive attitude and feedback received, subsequent meetings were organized with smaller differentiated groups. This allowed for a more practical and elaborate discussion with specific groups focusing on their potential involvement in this process, and ways in which such discussions could materialize locally. Most importantly, such meetings were particularly helpful in establishing relations between the local actors themselves, such was the example of local businesses operating in the field of accommodation, restaurants, and local producers of dairy products, bakery, and honey. To best illustrate the importance of connections among local businesses, and the need to develop into a *value chain,* a simple experiment was conducted "the breakfast experiment" (see Fig. 10.3). Two breakfast typologies were prepared: (a) a simplified version of the continental breakfast; (b) a home-made breakfast consisting of dough-nuts, honey, jam, and locally grown tea. Participants to the meeting were asked to choose which breakfast they thought tourists would prefer. The majority of participants opted for the continental version, arguing that a tourist would expect no less than a good *croissant for breakfast!* When told that, the home-made version was in fact a metaphor for the local value chain served all on one single plate, participants were surprised.

Through this simple experiment alone, one could easily identify a value chain consisting of several producers: the accommodation provider (including the restaurant), the honey producers, the farmer cultivating and collecting the local tea, the local grain producers, the local diary producers, as well as other suppliers of basic ingredients including packaging. From this comparison, the local businesses were introduced to the importance of enhancing the unique experience of a tourist with local flavours, establishing partnerships between them.

Alongside with the work conducted with local actors, the team worked on mapping and mobilizing interest among tour operators and tour guides operating nationally. They were contacted individually and introduced to the diverse natural resources and itineraries Gramsh had to offer. They were encouraged to introduce Gramsh to their networks and touring programmes and in exchange they were offered assistance and intermediation to establishing partnerships with local services providers in Gramsh. To kick-start this process, as well as to promote the destination in tangible terms, "Natyra Fest—EjaNëGramsh" hallmark event was held in May 2017 for the first time.

The Municipality of Gramsh, in cooperation with the technical NGO, the university and a number of tour operators, and local businesses organized an outdoor festival on the last weekend of May 2017, inviting all nature enthusiasts to become part

Fig. 10.3 The Gramsh breakfast experiment (photo: A. Ciro)

of it. The event aimed at: promoting Gramsh as a tourist destination particularly in outdoor active tourism—hiking, trekking, camping, rock climbing, cycling, caving, and canyoning; showcasing the potential of Gramsh as a tourist destination to the local businesses and the community; stimulate the local community to develop economic activities in accordance with the basic principles of sustainable tourism.

Agreeing on a commonly shared vision. Having kindled interest among the various actors involved, and a good understanding of the intricacies, the process progressed towards the development of a commonly shared long-term development vision for the municipality. Significant input from the various research, consultation processes and the hallmark event, helped identify gaps and weaknesses in service provision, as well as test the local capacities in handling a potential increase in tourist inflows. The latter was particularly helpful in identifying ways to improve and expand the existing capacities to cope with eventual higher tourist demand. The agreed vision oriented the municipality's long-term development towards sustainable tourism development, specifically building on its agricultural profile.

Orienting the economic activity towards a shared vision. It was the three parties' concern and key driving principle that, any strategic documents and processes undertaken would need to be as applicable and relevant to the context, as possible, so to avoid the standard-formal approach, with most documents ending up unused. To this end, and in compliance with the strategic planning approach, all the local actors together with the three parties leading the process, worked on translating the vision into strategic objectives, each detailed in investment projects (see Fig. 10.4). As part of the process, a number of local businesses were encouraged to align their objectives and long-term plans with the overall tourism development goal. The investment priorities reflected by and large the findings from the field work conducted, particu-

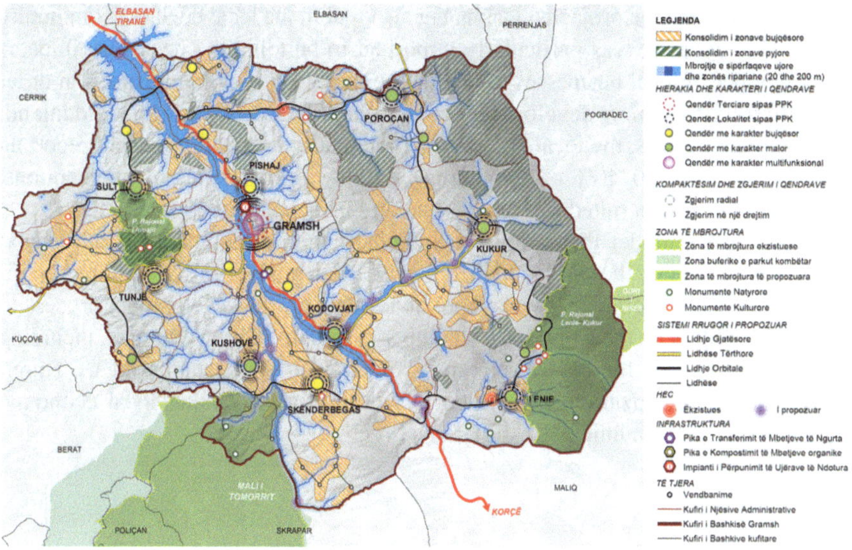

Fig. 10.4 Map of the vision—Municipality of Gramsh (source: Co-PLAN, Institute for Habitat Development)

larly the lack of quality infrastructure connecting the constituent administrative units of the municipality. Such investments would require significant amounts of financial resources, which cannot be covered by the municipal budget alone. Hence, shortage of financing means meant that, a good part of the investment priorities would have to rely on external financing, such as donor agencies assistance.

Improving and expanding the existing tourist supply. The sixth step of the model focused on the improvement and expanding of the existing tourist supply. The Municipality of Gramsh and the local business community lacked prior experience in tourism destination management, both a drawback and an advantage at the same time branding Gramsh as a destination to be explored. It was therefore important to "test" their tourist hosting and managing capacities and assess how local supply could be further improved and expanded. The hallmark event "Natyra Fest" EjaNëGramsh 2017 did precisely that, providing local business communities with an interactive assessment and tangible feedback on the quality of the products and services offered. Some businesses were also offered support in improving the design and promotion of their services and products. Over a weekend, a total of five national tour operators offered outdoor activities in Gramsh including hiking, canyon exploring, cycling, sight-seeing, bringing at least 550 tourists to the city, a good part of whom spending the weekend in Gramsh. For the local actors, this interest was a positive surprise.

The application of the model was important not only from a development perspective. The phasing of the model in steps allowed for the engagement of the local actors in participatory processes, which in turn built trust not only vis-à-vis the process and its legitimacy, and the projected vision, but also within the local business community itself. This dynamic was particularly important to building the necessary impetus for a number of local businesses to restructure their way of doing business in order to realign with the new development context and the evolving touristic demand. In quantitative terms, the local businesses, particularly hospitality related, report an increase of about 30–40% of their turnover following the series of interventions, resulting in newly recruited staff and investment in expanding their capacities.

More than a model that would "accomplish" the development of the context within its life cycle, the model applied served as an enabler of the required confidence and business climate, two key pre-conditions for the development process to take off. Building on the work done, a number of initiatives have emerged, including a dedicated Center for the Development of Gramsh was established as a non-governmental organization, committed to working on catalysing local economic development in the municipality of Gramsh.

10.5 Conclusions

The case study of Gramsh showcases that sustainable tourism is possible even in peripheral territories with scarce resources. The pre-conditions to that however are: (1) the engagement of the local community including the local businesses in decision-making processes as to how the local natural resources will be used; (2) building and sharing confidence in the process; (3) mobilizing partnerships locally and beyond.

The use of bottom-up processes for tourism innovation strengthens the efficient use of endogenous development potentials. It is in the direct interest of the local communities that the natural resources are used in the most considerate and far-sighted way possible since their long-term welfare depends on them. Hence community participation from an early stage in outlining a development vision for the municipality, and how these resources will be used to that effect, is imperative to ensure the effectiveness and longevity of the process. Such an approach boosts local identity and self-identification of citizens with own territory and local resources. It builds trust and confidence in the process, which can then yield partnerships essential to the establishment of a value chain.

The case study of Gramsh shows that external support is required, as the local capacities to organize tourism offers and to do a minimum of tourism destination management, are not developed and communities do not know well how to get started. This holds particularly true when it comes to place-branding, which is where a "destination management organization" could play an important role.

The local community also does not recognize that the concept of sustainable tourism, respecting the environment and the community, can be applied for unique tourism experiences, with visitors getting in touch with the local community rather than relaxing in comfortable but impersonal sun-sea-sun resort. In this regard, the process in Gramsh has convinced the local community that they have something to offer. The community has also seen which kinds of activities are needed to attract tourists—and how important the development of networks is. The process described in this chapter has ended with the Nature Fest and has discussed the preparatory actions and the first events to establish Gramsh municipality as a tourism destination. A critical phase is coming now, which is whether the community has the organization capability to carry on with the activities started. The municipality and the new local NGO will have an important role in this regard.

The case study of Gramshcostacan be further elaborated and be developed into a replicable development model, as the challenges other local governments are faced with are not very different from those in Gramsh. Rapidly changing contexts, demanding local constituencies for better service provision, increasing competitiveness among regions are pushing municipalities to find innovative solutions to reinvent themselves in ways that translate into local economic development.

References

Aall C (2014) Sustainable tourism in practice: promoting or perverting the quest for a sustainable development? Sustainability 6(5):2562–2583

Angelevska-Najdeska K (2012) Planning of sustainable tourism development. Procedia Soc Behav Sci 44:2010–2020

Beyer M (2014) Tourism planning in development cooperation: a handbook. Deutsche Gesellschaft für Internationale Zusammenarbeit (GIZ), Bonn

Boka M, Torluccio G (2013) Informal economy in Albania. Acad J Interdiscip Stud 2(8):212–221

Conway S, Steward F (2009) Managing and shaping innovation. Oxford University Press, Oxford

Costa C, Brandão F (2011) Tourism Innovation through effective public participation: a sustainable approach. In: International Tourism Sustainability Conference 2011 "Embracing social and environmental change: the influence and role of tourism" 21–24 September. University of Technology of Mauritius

European Union (2006) Sustainable tourism as a factor of cohesion among European regions. CoR Studies, Brussels E-6/2006

Getz D (2008) Event tourism: definition, evolution and research. Tour Manag 29:403–428

Hjalager AM (2010) A review of innovation research in tourism. Tour Manag 31(1):1–12

Kline S, Rosenberg N (1986) An overview of innovation. In: Landau R, Rosenberg N (eds) The positive sum strategy: harnessing technology for economic growth. National Academy Press, Washington, DC, pp 275–304

McKercher B, Prideaux B (2014) Academic myths of tourism. Ann Tour Res 46:16–28

Medina L, Schneider F (2018) Shadow economies around the world: what did we learn over the last 20 years? International Monetary Fund Working Paper

Moulaert F, MacCallum D, Mehmood A, Hamdouch H (2013) The international handbook on social innovation. Collective action, social learning and transdisciplinary research. Cheltenham, Edward Elgar Pub

Nientied P, Karafili E (2016) Towards a pragmatic perspective on business innovation in Western Balkan countries: the case of Albania. Int J Business Manag 3(11):193–202

Nientied P, Ciro A, Porfido E (2018) Sustainable tourism development in Albania in times of liquid modernity. In: Proceedings international workshop: enhancing sustainable tourism in Adriatic-Ionian region through co-creation. Macerata, Italy, 15–16 September 2016

Pan B, Crotts JC (2012) Theoretical models of social media. In: Sigala M, Christou E, Gretzel U (eds) Social media in travel, tourism and hospitality: theory, practice and cases. Ashgate, Surrey, pp 73–86

Phills J, Deiglmeier K, Miller DT (2008) Rediscovering social innovation. Stanf Soc Innov Rev 6(4):34–43

Rodríguez A (2009) Social innovation for neighbourhood revitalization: a case of empowered participation and integrative dynamics in Spain. In: MacCallum D, Moulaert F, Hillier J, Vicari Haddock S (eds) Social innovation and territorial development. Ashgate, Farnham, pp 81–100

Singh T (ed) (2017) Challenges in tourism research. Channel View Publications, Clevedon

Tidd J, Bessant J (2013) Managing innovation: integrating technological, market and organizational change. Wiley, Chichester

Tony D, Epstein M, Shelton R (2012) Making innovation work, how to manage it, measure it, and profit from it. Pearson FT Press, Upper Saddle River

UNWTO (2013) Sustainable tourism for development guidebook. UNWTO, Madrid

Van Dyck B, Van den Broeck P (2013) Social innovation: a territorial process. In: Moulaert F, MacCallum D, Mehmood A, Abdelillah A (eds) The international handbook on social innovation. Collective action, social learning, and transdisciplinary research. Edward Elgar, Cheltenham, pp 131–141

World Travel & Tourism Council (2016) Travel & tourism economic impact 2017. London

World Travel & Tourism Council (2017) Travel & tourism economic impact 2017. London

Chapter 11
Professional Capacity Building as the Public Sector Intervention Towards Sustainable Economic and Territorial Development: Key Study Knowledge Hubs Network for Energy Efficiency

Dragomir Tzanev

11.1 Introduction

The chapter is focused on the public sector role in professional capacity building necessary for the improvement of energy efficiency in the building sector at the national and local level, which has direct implications on the territorial and urban planning and the design of the national strategic documents and programmes targeted to the residential buildings. It brings together the results achieved by four consecutive EU-financed projects in reforming of the outdated training practices and implementation of nearly zero-energy building (NZEB) in Bulgaria, and consequently, in other countries in Eastern and Southern Europe. This chapter shows the legacy of the BUILD UP Skills initiative (European Commission 2018a) as public interventions towards higher energy efficiency of the built environment via improvement of the skills in the construction sector.

Although a number of barriers are still left, the idea for establishing of a network of well-equipped training centres (Building Knowledge Hubs) has proven a great potential to increase the capacity to deliver new buildings and renovations with the highest possible quality, and thus, to strongly influence the urban development—especially in specific Eastern European countries, where most of the interventions in the building sector are motivated by the availability of public supporting schemes. As there are still many challenges on the road ahead—mainly in terms of qualified trainers, up-to-date training facilities and local capacity building for the professional communities, after a common EU surge through the two-staged BUILD UP Skills initiative, the idea gained further support by Horizon 2020. Five new such centres (in Bulgaria, Romania, Czech Republic, Turkey and Ukraine) are established, all of

D. Tzanev (✉)
EnEffect, Center for Energy Efficiency, Sofia, Bulgaria
e-mail: dtzanev@eneffect.bg

© Springer International Publishing AG, part of Springer Nature 2019

M. Finka et al. (eds.), *The Role of Public Sector in Local Economic and Territorial Development*, EAI/Springer Innovations in Communication and Computing, https://doi.org/10.1007/978-3-319-93575-1_11

them deliver training courses on specific energy efficiency related programmes and providing consultations for various stakeholders' groups, including local authorities. Three new centres (in Greece, Croatia and Italy) have joined the network through another project, Fit-to-NZEB, looking to promote training schemes targeted to deep energy renovation and influence political decisions at national and local level supporting the long-term goal for decarbonizing the European building stock.

11.2 Concept and Approach

The building sector plays an important role for the execution not only of the 2020 climate and energy package, but also for the years ahead, as it still represents about 40% of energy use in the EU with the largest cost-effective savings potential (European Commission 2018b). In the view of the debates on the energy and climate goals for 2030 and 2050 and in the light of the Energy Union initiation and the issues with the security of the energy supplies, the EU has adopted an ambitious vision for the energy performance of its buildings as well. It is well-known that by 2020 all new buildings should be nearly zero-energy buildings, but the real challenge is in the Member States, which should draw up ambitious action plans for increasing the numbers of NZEBs, including policies and measures for refurbishment of existing buildings. A major transformation will occur in the building sector, which represents a major challenge to the construction industry. Respectively, equally major efforts are needed to improve the skills around the whole value chain in the sector: investors, end users, architects, civil engineers, building managers, construction workers and other building professionals, and notably, national and local authorities, emphasizing on the importance of integrated building design approach within a coherent urban development concept for execution of quality NZEB projects.

The economic triggers are even more stringent: even in times of severe financial crisis, when the BUILD UP Skills initiative stared, the EU-28 construction sector accounted for more than 5% of (gross) value added on average, reaching more than 10% in Romania (EUROSTAT 2017). In Bulgaria, the construction industry employs about 6% of the working force and is the biggest industrial employer in the country (Bulgarian Construction Chamber 2017). Although it has been one of sectors worst hit by the economic crisis, most countries report only a slight shortage of building workers and specialists by 2020, but the need for re-training of the current professional workforce is much stronger and is recognized throughout Europe and in particular in countries in Eastern Europe. This highlights the importance of continuing education for the current workforce. To support such efforts, in addition to the European "Agenda for new skills and jobs" (European Commission 2010) and "New Skills for New Jobs" initiative (Commission of the European Communities 2008), in 2012 a Communication on a "Strategy for the sustainable competitiveness of the construction sector and its enterprises" (European Commission 2012) was published, focusing on the human capital in the sector. The Energy Efficiency Plan of the Commission (European Commission 2011) also recognizes the importance of training and urges Member States to develop strategies and schemes to respond to the training needs, and this stance is strengthened by specific texts in Directive 2010/31/EU on the Energy Performance of Buildings (Art. 20, Art. 17) (European Parliament and Council of EU 2010) and Directive 2012/27/EU on Energy Efficiency (Arts. 8, 16, 17) (European Parliament and Council of EU 2012).

Thus, reaching of the 2020 energy and climate change objectives requires a major quality shift throughout the construction sector, which needs to be ready to deliver high energy performing renovations and, in particular, nearly zero-energy buildings. This goal necessitates a major effort to increase the number of qualified professionals and building workers, which is directly related to the quality of the educational programmes and the inclusion of training on intelligent solutions for energy efficiency and renewable energy in buildings. As long-term impacts, also corresponding to the latest market demands, these changes are expected to contribute to the overall execution of the 2030 and 2050 goals (exemplified by the "Clean Energy for All Europeans" legislative package (European Commission 2016)) in achieving of decarbonized building stock, integration of decentralized energy production based on renewable sources and, as an overarching goal, climate change mitigation: targets which achievement is virtually impossible without the support of coherent urban development policies.

Of course, a number of barriers exist, especially in markets driven by low-cost investment decision as in many Eastern European countries. Some of these barriers include the fragmentation of the labour market and regional discrepancies, lack of understanding of the NZEB specifics by the professional community, low motivation of the workforce and low educational level of the blue-collar workers; lack of experience and practice with nearly zero-energy renovations, low level of penetration of RES technologies, lack of cooperation between education providers, local authorities and business actors, and many others. There is a pressing need for upgrading and facilitating the available large-scale qualification and training schemes and also to promote flexible continuing training schemes for architects, engineers, building managers, local authorities and other influencers, emphasizing on the integrated building design approach for execution of quality NZEB projects. Thus, through the establishment of a well-designed networking structure supported by enhanced web-based communication tools, the process will involve all relevant target groups, focusing not only on actors identified in the BUILD UP Skills initiative, but also targeting the existing European city networks working in the area of sustainable energy development. Last but not least, the deployment of a targeted information and communication campaign should lead to the attraction of some of the most important and often underestimated actors in the field of NZEB project implementation—the end users and the investors, also central to the urban planning process. This should be supported by networks of one-stop shop training, information and consultation centres for implementation of NZEB projects, providing not only integrated building design services and demonstration of various solutions related to the improved energy efficiency of the projects, but also administrative and financial advice (extremely important in situations, in which building projects rely to a large extent on EU cohesion policy funding). Thus, the difficulties to gather all necessary expertise and skills for the quality execution of any building project—and especially of a NZEB—could be successfully overcome, which will remove one of the major barriers in satisfying the increasing public demand for energy efficient buildings with improved living comfort and economic performance.

11.3 Achieved Results

Step 1: BUILD UP Skills Pillar 1: National consultation platform, Roadmap for improving the skills in the construction sector and widespread endorsement campaign.

Starting in 2011 with identical projects in 30 European countries, the BUILD UP Skills initiative responded to the recognition that reaching the 2020 EU energy and climate goal represents a major challenge to the construction sector, requiring a major effort to increase the number of qualified construction specialists at all levels, which is directly related to the accessibility and quality of the educational programmes. Since November 2011, the project "Roadmap for Trainings to Develop Skills and Knowledge on Intelligent Energy Solutions in Buildings for Bulgaria until 2020" (BUILD UP Skills Bulgaria), prepared by a consortium consisting of Center for Energy Efficiency EnEffect (coordinator), National Agency for Vocational Education and Training and Bulgarian Construction Chamber, entered its implementation phase in an effort to unite all stakeholders involved in the construction sector and the national education system to dramatically increase the knowledge and skills for construction of nearly zero-energy buildings.

The National Platform for Dialogue, which was initiated in the project, provided the foundation of a comprehensive process of consultation and cooperation of public authorities in the field of education and construction, professional associations, chambers of commerce and industry, associations of producers of building components and materials and separate business actors, professional high schools and vocational training centres, energy suppliers and distributors, NGOs working in the areas of energy, climate and environment protection. Within the project, a detailed study of the current situation was conducted, including:

1. Analysis and evaluation of the situation in the construction sector, including the latest statistical data.
2. Analysis and evaluation of national policies to achieve the objectives of the Energy Strategy of the Republic of Bulgaria until 2020 (Ministry of Energy of Republic of Bulgaria 2011) (including the expected contribution of the construction sector).
3. Analysis and evaluation of the national system for continuing education in the construction sector and the extent to which it meets the needs of additional skills for the implementation of energy efficiency measures and installations for RES in buildings.
4. Identification of the existing barriers and gaps in vocational training in the construction sector; measurable data for the needs of skilled workforce by 2020.

Based on this study, a National roadmap and Action plan to develop skills and knowledge on intelligent energy solutions in buildings for Bulgaria until 2020 has been prepared and was endorsed by more than 20 responsible authorities and stakeholders, with the commitment to implement and enforce the set of concrete measures and targets. It identified 6 priority areas:

- Priority 1. Measures for upgrading of the professional skills
- Priority 2. Training of trainers: new or upgraded systems for qualification and requalification
- Priority 3. Improvement of the training facilities
- Priority 4. Structural measures for monitoring of the processes
- Priority 5. Interaction among the stakeholders
- Priority 6. Raising the citizens' awareness about the opportunities for energy efficient solutions

Step 2: BUILD UP Skills Pillar 2: New training programmes, training of trainers, pilot courses and interactions among stakeholders.

Actively supported by the Germany-based Passive House Institute, an extended project team joined by active education and training providers initiated the second stage of the BUILD UP Skills initiative in Bulgaria; as a result, significant changes were initiated in the vocational education and training system, integrating energy efficiency concepts into the national educational plans and programmes for professional high schools of construction and architecture. This, of course, did not come without sufficient preparation work. In the first stages of the project, a Centre of Excellence was established as an open structure focused around the consortium members. It undertook the work on gathering and analysing of the required technical competences according to the identified needs, which had to be included in the training programmes throughout national training system and in the newly developed training programmes. Two teams of experts were selected to perform the review of the existing training materials and to develop the general training schemes under which specific training programmes were elaborated and future trainers were trained in the targeted professional directions—"Construction" and "Electrical Engineering". In parallel, the existing state educational standards were reviewed, with specific proposals for updates based on the conclusions and the input from the Centre of Excellence.

In the next stage of the project, a team of highly qualified specialists was selected and completed extensive training as future trainers. It included undergoing a brand-new e-learning training module set up by PHI in the course of the project, currently available for use all over the world at https://elearning.passivehouse.com/, also emphasizing on the importance of urban planning approaches for achievement of ambitious energy efficiency standards. Then, the trainers completed a specialized training in Passive House Academy—Dublin, Ireland, designed according to their specific needs and requirements. After the training programmes were finalized, train-the-trainer courses were conducted in Bulgaria, covering seven cities and more than 100 trainers. Most of the training sessions were attended by representatives of the local authorities due to the relations established with the Municipal Energy Efficiency Network EcoEnergy.

As next steps, the new ten training programmes for construction workers were finalized and all relevant training materials, aids and supporting tools were developed. This allowed the execution of actual trainings by each of the involved training institutions, covering more than 300 construction specialists and workers. The trainings were closely monitored with the goal to ensure constant quality checks and improvement of the training schemes and programmes.

These activities were accompanied by a systematic and active dissemination and communication campaign, designed to support the engagement and confirm commitment of key actors and promote new opportunities through increased visibility of the benefits of the improved training schemes. The execution of the communication strategy provided the transparency of the activities and the results of the project, succeeding to motivate and engage the leading actors and stakeholders and attract participants to the training courses.

The results are more impressive—both trainees and construction companies are extremely pleased with the quality of the courses and the scope of the acquired knowledge, skills and competencies. Based on the newly developed programmes and the experience from the conducted training courses, the National Agency for Vocational Education and Training (NAVET) officially approved a plan to update the state educational standards for acquiring qualification for 13 professions in professional directions 522 "Electrical and Energy" and Professional 582 "Construction". Additionally, a programme for a new subject in the professional high schools of architecture and civil engineering called "Energy efficiency in construction" was developed by the project partners.

Step 3: Train-to-NZEB: New training facilities, diversified training offer and one-stop-shop consultation centres.

However, it was clear that there are a number of barriers still left and the lack of adequate training facilities to conduct practical trainings represented one of them not only for Bulgaria, but also for many other countries in Eastern and Southern Europe. Thus, the idea for establishing of a network of well-equipped training centres (Building Knowledge Hubs, BKHs) was developed and gained further financial support by the Horizon 2020 of the EU in the project Train-to-NZEB (www.train-to-nzeb.com). Five new such centres—in Bulgaria, Romania, Czech Republic, Turkey and Ukraine—were established under the guidance of PHI and MosArt/Passive House Academy (Ireland), as currently most of them deliver training courses under the BUILD UP Skills curricula and international certification schemes like the Certified Passive House Designer/Consultant and Tradesperson programmes. Additionally, a number of other training programmes are developed in close contact with the national authorities responsible for the vocational training sector, the professional chambers, the industry and other vocational training providers, which diversifies the training offer and channels new trainees to the Passive House training schemes. The centres have proven themselves as a valuable source for a range of new opportunities and educational approaches for the universities, as for example the project supported the development of three disciplines in the framework of a new Masters programme in the University of Architecture, Civil Engineering and Geodesy in Sofia, Bulgaria; stale connections and cooperation activities are also established with architectural and technical universities in Kyiv (Ukraine) and Izmir (Turkey)—where the local BKHs are situated, and in Bucharest (Romania) and Prague (the Czech Republic), jointly participating in training courses and professional events.

The BKHs are equipped with all necessary tools and materials to demonstrate in practice the leading principles of the NZEB concept and approach based on the "energy efficiency first" principle, with all of them having full pressurization testing and infrared imaging equipment, operational mechanical ventilation with heat recovery systems, practice training walls for application of airtightness materials and products, and full-size demonstration models showing the approach to the NZEB concept in different construction types specific for the participating countries. Through the integration of practical training exercises in the generally theoretical training schemes, a large number of industrial partners were attracted to support the development of the training centres and now use the facilities for training not only of construction workers, but of sales and marketing personnel, which is another positive feature of the undertaking. Additionally, the training programmes have proven to be a matter of interest (as according to expectations) for building managers and supervisors, who are best positioned to transfer the knowledge at the building sites, especially in countries with depleted human capital and largely underqualified working force in the construction sector.

Most importantly, being positioned as NZEB knowledge centres, the BKHs have attracted a number of enquiries for direct consultations on actual PH construction projects and have conducted series of training courses and demonstration sessions for local authorities. Supported by the strong international movement towards energy efficiency and climate change adaptation planning at city level and the national programmes supporting energy efficiency building renovation, the BKHs have proven to be a logical influencer of decisions for managing and strategic planning of development of the building stock. The cooperation with national energy efficiency networks of local authorities, as for example the continuing cooperation with the EcoEnergy network in Bulgaria, has led to considerable volume of activities not only in direct training of the municipal staff but also in consultations on energy efficiency action plans related to the building sector following the requirements of the national legislation and international initiatives like the Covenant of Mayors. Additionally, some of the partners managing the BKH have become a part of the national consultation platforms of strategic documents and financing programmes as the National Housing Strategy and the National Programme for Energy Efficiency of the Multifamily Residential Buildings, which have direct impact on the improvement of the urban environment at local level.

To promote these consultation activities, within the Train-to-NZEB project, a free e-learning tool specifically designed for decision makers with no professional competences was developed. A smartphone application is also under way, demonstrating the basic NZEB principles, solutions and details to investors, decision makers, construction specialists and workers—and, in fact, to anyone interested.

Next Steps

Three new centres (in Greece, Croatia and Italy) have joined the BKH network through another Horizon 2020 project, Fit-to-NZEB (www.fit-to-nzeb.com), looking to promote training targeted to the achievement of deep energy renovation, including through the step-by-step renovation approach. The project aims to develop training and educational programmes along all relevant levels of the European Qualification Framework (levels 2–7) (Cedefop 2018), including programmes for validation of knowledge and skills acquired at the workplace and continuous training for adult workers. A set of common learning outcomes is already elaborated, including specific topics of importance for the urban planning as optimal solar gains and integration of RES at district/neighbourhood level, among many others. New demonstration and practice training models will be developed, specifically presenting solutions suitable for deep energy retrofit and integration of renewables. The training programmes together with a number of training aids and tools will be available for the vocational training institutions around Europe, thus raising the interest and supporting the integration of the NZEB concept in the national educational systems. A special focus of the project is the involvement of decision makers at national and local level in the process of development of the programmes and in the pilot training courses, which will lead to better understanding of the topical issues and increased capacity to enable and stimulate NZEB solutions through urban planning and spatial development decisions.

11.4 Conclusions

Professional capacity building in the area of energy efficiency and RES in buildings offers a great instrument for public sector intervention towards delivery of quality nearly zero-energy buildings and increase of the building renovation rate, thus promoting sustainable economic growth within a holistic urban development perspective. The overarching goal of decarbonizing the European building stock by 2050 will streamline even more active public support for deep energy renovation and nearly zero-energy buildings, as city planners and local professionals will play a key role for the effective utilization of the funds and attracting of private investment in the renovation of the building stock and, respectively, regeneration of the urban landscape. However, in many EU countries new local capacities still have to be built—mainly in terms of qualified trainers, up-to-date training facilities and importantly, new knowledge and skills within the local administration.

The idea of sharing experiences among specialized training centres has attracted continuous support by the EC and has already been successfully replicated in eight countries, proving its further transferability and possibilities for market application in different social and economic conditions and within varying building traditions, urban context and existing building stock. Supported by local authorities on many occasions, the growing network of training and demonstration centres is also providing services for policymakers, municipal experts and investors, enabling the mobilization of private investment in the renovation of the building stock and the generation of the urban environment, thus supporting the overall market development and the local economic growth.

References

Bulgarian Construction Chamber (2017) The Bulgarian Construction Industry in 2016. Annual analysis. http://ksb.bg/images/NOVO1/analizi/Presentation-2016-godishen.pdf. Accessed 16 Jan 2018

Cedefop (2018) European qualifications Framework (EQF). http://www.cedefop.europa.eu/en/events-and-projects/projects/european-qualifications-framework-eqf. Accessed 16 Jan 2018

Commission of the European Communities (2008) New skills for new jobs anticipating and matching labour market and skills needs. Communication from the Commission to the European Parliament, the Council, the European Economic and Social Committee and the Committee of the Regions. Brussels, COM (2008) 868/3

European Commission (2010) An agenda for new skills and jobs: a European contribution towards full employment. Communication from the Commission to the European Parliament, the Council, the European Economic and Social Committee and the Committee of the Regions. Strasbourg, 23.11.2010 COM (2010) 682 final

European Commission (2011) Energy Efficiency Plan 2011. Communication from the Commission to the European Parliament, the Council, the European Economic and Social Committee and the Committee of the Regions. Brussels, 8.3.2011 COM (2011) 109 final

European Commission (2012) Strategy for the sustainable competitiveness of the construction sector and its enterprises. Communication from the Commission to the European Parliament and the Council. COM/2012/0433 final

European Commission (2016) Clean energy for all Europeans. Communication from the Commission to the European Parliament, the Council, the European Economic and Social Committee, The Committee of the Regions and the European Investment Bank. http://eur-lex.europa.eu/resource.html?uri=cellar:fa6ea15b-b7b0-11e6-9e3c-01aa75ed71a1.0001.02/DOC_1&format=PDF

European Commission (2018a) BUILD UP Portal: Skills/About BUILD UP Skills. http://www.buildup.eu/en/skills. Accessed 16 Jan 2018

European Commission (2018b) Official website. https://ec.europa.eu/energy/en/topics/energy-efficiency/buildings. Accessed 16 Jan 2018

European Parliament and Council of EU (2010) Directive 2010/31/EU of the European Parliament and of the Council of 19 May 2010 on the energy performance of buildings. http://eur-lex.europa.eu/legal-content/EN/ALL/;ELX_SESSIONID=FZMjThLLzfxmmMCQGp2Y1s2d3TjwtD8QS3pqdkhXZbwqGwlgY9KN!2064651424?uri=CELEX:32010L0031

European Parliament and Council of EU (2012) Directive 2012/27/EU of the European Parliament and of the Council of 25 October 2012 on energy efficiency, amending Directives 2009/125/EC and 2010/30/EU and repealing Directives 2004/8/EC and 2006/32/EC Text with EEA relevance. http://eur-lex.europa.eu/legal-content/EN/TXT/?qid=1399375464230&uri=CELEX:32012L0027

EUROSTAT (2017) Construction production (volume) index overview. http://ec.europa.eu/eurostat/statistics-explained/index.php/Construction_production_(volume)_index_overview. Accessed 16 Jan 2018

Ministry of Energy of Republic of Bulgaria (2011) Energy strategy of the Republic of Bulgaria till 2020. https://www.me.government.bg/files/useruploads/files/epsp/23_energy_strategy2020%D0%95ng_.pdf. Accessed 16 Jan 2018

Chapter 12
The Public in Search of Identity: New Symbolism in Urban Spaces. A Study of Central Squares of Balkan Capitals

Peter Nientied and Besnik Aliaj

12.1 Introduction

In this chapter, we present the findings of a study on identity formation through the main urban squares of capitals of four small Western Balkan countries: Skopje (FYR of Macedonia; hereafter: Macedonia), Tirana (Albania), Pristina (Kosovo) and Podgorica (Montenegro). Our interest in this topic of national identity formation in space and place was raised during the rather dramatic redevelopment process of the main square of Skopje and the political discussions regarding the main square of Tirana. We decided to take a closer look at these developments and explore what the new squares were meant to express. We broadened the study and included the cases of main squares of two neighbouring capitals, Pristina and Podgorica. Their main squares did not reach the news often. We study how the squares have been subject to government intervention to renew the squares as element of national identity formation.

Identities are given shape in places, including landscapes, monuments, and sites where commemorations are performed. In these places, collective memory is reinforced, and national identity is constructed, both formally and informally (Osborne 2001). Main squares in capitals have a national symbolic value and are generally a focal point to articulate national identity. Capitals are essential symbols of national iconographies (Ter-Ghazaryan 2013) and their main squares are symbolic spaces. Main squares signify a place where people get together to remember or celebrate,

P. Nientied (✉)
Co-Plan Institute for Habitat Development, POLIS University, Tirana, Albania

Institute for Housing and Urban Development Studies, Erasmus University, Rotterdam, Netherlands
e-mail: nientied1@chello.nl

B. Aliaj
Polis University, Tirana, Albania
e-mail: besnik_aliaj@universitetipolis.org

© Springer International Publishing AG, part of Springer Nature 2019
M. Finka et al. (eds.), *The Role of Public Sector in Local Economic and Territorial Development*, EAI/Springer Innovations in Communication and Computing, https://doi.org/10.1007/978-3-319-93575-1_12

be it a tribute of an historical event or the reception of a sport team winning a special medal. Main squares are principal places for political demonstrations and are also ordinary spaces—places to meet and spend some time.

The four squares studied are vital public spaces in the country from the perspective of national symbolism. Meanings related to material objects of the built environment are studied, and as such this study can be seen as a study in urban semiotics (Sepe and Pitt 2014). Indeed, in cases like these four central urban squares, post-socialist national identities are at stake (Light and Young 2013). In Tirana and especially Skopje, interventions have been very substantial and have taken considerable financial and immaterial costs. The cases of Podgorica and Pristina look to be more unpretentious but are relevant. *Tirana's* central square has been the place for all kinds of manifestations, including protests against governments. Different governments have sought to redesign the square, creating a national symbol with various cultural connotations. After a redevelopment, a new square was inaugurated in June 2017. The case of *Skopje's* central square is extraordinary. After the earthquake of 1963 that hit Skopje, the city was reconstructed, in a modernist fashion. The government decided in 2008 to redevelop the inner city, create a new square and give expression to a new national narrative on the origins of Macedonia. Much money was spent on the construction of museums and government buildings, as well as the erection of monuments depicting historical figures from the region of Macedonia. Fierce urban and political struggles have accompanied this process. *Pristina*'s central square has been relatively uncontested—unlike the political processes in this young country. As 'square' we take two squares and a boulevard that together have the function of a central square. The square contains the first symbols of the country's new and still questioned independence. In *Podgorica*, Montenegro, also a new country that came into existence in 2006, the main urban square was renamed in 2006 to Republic Square and was improved, as an element of a bigger inner-city renovation, to westernize the city and forget much of the past. The square was renamed and again renamed in 2016 to Independence Square. The name—not the square itself—is a symbol in the continuous political struggle in the country.

Before discussing the main squares and the governments' interventions, first a brief sketch is given of the context of Central and Eastern European (CEE) urban transformation, and then attention turns to the concept of identity and the theoretical framework employed in this chapter.

12.2 CEE Urban Transformation

After 1990, CEE countries went through a period of deep transformation from socialist political systems to democratic systems and from state organized economic systems to more market led economic systems. The period of change—which is still going on—has been demanding, for citizens and politicians, for institutions and companies. Hopes were high during the early 1990s. Of course, there was in these days no experience with and no clear view of how economic, political, social and cultural processes would develop after the change of system. Countries like Spain, Portugal and Greece also went through deep changes of regime, from authoritarian to more democratic systems. But the changes in CEE countries were more than political changes and comprised economic change from state socialism to market-led economies, and cultural change from relative isolation to global economic and cultural exposure and relationships. Several countries had to create new national identities (e.g. from former Czechoslovakia and Yugoslavia). Political and institutional maturity in the Balkan countries was far less than in countries like Greece, Portugal and Spain. Bollens made a comparative analysis of public policies in cities in Spain and Bosnia, to conclude that 'While the Spanish cases exhibit actualized potential in how urban interventions can affect positive change, the Bosnian cases reveal missed opportunities' (Bollens 2009, p. 102).

Post-socialist cities have changed substantially—and are still in a process of transformation. The term 'post-socialist' is used on purpose, notwithstanding pertinent questions that authors like Ferenčuhová (2012) and Tuvikene (2016) have raised. A point they raise is: the changes took place more than 25 years ago—is it still relevant to typify Central and Eastern European cities based on their socialist past? For this study, it is. In the cities of Tirana, Skopje, Podgorica and Pristina, the socialist past is very visible, not only in physical terms, but also in cultures and mindsets. If we were to study say tourism, we would not use the label 'post-socialist tourism'. But in cities, artefacts show history, collective memories and identity are based on history. As Sýkora and Bouzarovski state '...post-communist cities are subject to three aspects of post-communist transition: institutional transformations, transformations of social practices and, transformations in urban space. While the formal remodelling of the institutional landscape has now been largely completed in many former communist countries, social practices and structures still retain some socialist features and large parts of post-communist cities exhibit a typically socialist urban character. Therefore, we have argued that post-communist cities are still cities in transition' (Sýkora and Bouzarovski 2012, p. 53).

Most CEE cities had ample public space during communist times. 'Consistent with Soviet theory on the importance of public space for the construction of a new "socialist man," during the period of state socialism cities in these countries were endowed with massive public spaces, such as parks and plazas, whose purpose was ceremonial and disciplinary than democratic' writes Hirt (2014, p. 124). In the new political and economic realities after 1990, most cities—including the case study cities—witnessed a decrease of public space, due to m arket sector led infill. Local governments issued construction permits without giving much attention to the consequences like substantial densification (Mariotti and Koželj 2016; Pojani and Maci 2015) and infrastructure and mobility problems.

Building new political and cultural identities was not on top of the formal political agenda during the early phase of the transition. Sýkora and Bouzarovski (2012) point out the multiple transformation dynamics of institutional, social and urban transitions. These multiple transitions follow a specific temporal sequence. In general, they move from addressing universal to more specific areas. Urban transformations would not happen without transformations in the basic political and economic rules.

Not everywhere in CEE this sequence was witnessed: during the early 1990s, in Tirana for example, a process of massive informal city development took place (Aliaj et al. 2003). Cities throwing off old legacies and their socialist identities have received attention from practitioners and academicians (Paskaleva and Cooper 2017). 'The disappearance of the material capital of the socialist past has been accompanied by intense commemoration practices verging on memorial obsessions. New monuments, plaques, street names, and museums appeared almost as quickly as the old "disposable" ones were forced out from the urban landscape.' says Lisiak (2009, p. 431; cf. Czepczyński 2010; Begić and Mraović 2014). The 'away from socialism' was one step, but carefully fostering a new national identity is quite another. Light and Young (2015) suggest that throughout CEE and the Soviet Union the exit from socialism was marked by an immediate and public rejection of socialism. Socialist regimes had enjoyed little popular support or legitimacy in their final years among a populace eager for change. Post-socialist governments sought to establish new political identities as liberal European states which deserved European support and with which the West could do business. This resulted in 'landscape cleansing'. In the decades after 1990, these beliefs and attitudes developed in multiple ways. Recently post-socialist societies have developed a more ambivalent approach to their recent past. In Tirana, recently two museums have been opened showing the 'realities' of the socialist past (the history displayed is of course disputed), for cultural and economic reasons. Tourists appreciate it, like they are keen to visit the Stasi-museum in Berlin.

In post-socialist cities, attacking and pulling down symbols of the socialist state was amongst the most visible and convincing expressions of the transformation from socialism to new political systems. The fall of the Berlin Wall was a major occasion, globally symbolizing the end of an era. Statues of former communist heroes were objects with high symbolic value. Statues may have gone done, but most artefacts that have been built during the socialist period have been kept of course (Lisiak 2009), for various reasons. Parks may have had symbolic values, but are appreciated too in post-socialist times, are not seen as socialist symbols, especially after a name change from say Lenin Park to something more modern-day. Housing and planning are of course needed in the post-socialist period too. Many buildings with a public function could be adjusted to the new system (e.g. a government office or a university). And some buildings were just too complicated and/or disputed to be destructed. Perhaps the most dramatic example is the Casa Populori in Bucharest—which is a hated symbol of the Ceausescu socialist period but was too big to demolish (Light and Young 2013).

12.3 Identity and Main Squares of Capitals

CEE states were confronted with questions about their national identity. National identities are historical constructions that are constantly being reconstituted according to a presentist agenda (Osborne 2001), i.e. viewing the past with a somewhat limited to present-day set of attitudes and beliefs. In this study, we deal with *place-related identity*. In social sciences, this concept has various interpretations, such as sense of place, place attachment, urban identity, and so on. Articles of Belange et al. (2017), Cheshmehzangi (2011, 2015) and Lewicka (2011) give overviews of theoretical foundations of the concept of place-related identity. Urban or place identity has two sides; it can be seen as a feature of a city based on a collective attribution, and as self-identification of a person with the city. Cheshmehzangi (2015) states that the identity of a city or a place needs to be attached to at least one of the fundamental factors of history, meaning, distinctive characteristics and purpose. These factors can be connected to main urban squares, as we will see in the case studies. In our study of main squares of capitals of countries, identity is not just 'local identity', a revised national identity is expressed at a local level.

A city's main square can be understood as a type of urban landscape, and this can be understood not only as a built morphology of the city—its urban tissue, outlook and layout—but also as a complex set of meanings, representations and values which it conveys, and in which has been invested (Murzyn-Kupisz and Gwosdz 2011). Places, such as urban squares, have no inherent identity. Identity is a mental quality, it is constructed by human behaviour in reaction to places (Osborne 2001, p. 4). It is a reciprocal relationship—people produce places and they derive identities from them. Identity may be given to a place by a government, but whether this identity is embraced by the people is a different matter. Places are both material and social, in the sense that they provide space people use for all kinds of activities. Celebrating the victory of a team, organizing an urban protest, a president speaking to a crowd, etc., all signify identity activities of groups, between groups, and institutions like government. Various groups and individuals may have different associations with a specific place: The Red Square in Moscow is a beloved icon for one group, a disliked symbol of domination for another, or a striking tourist spot for again another.

Main squares in capitals can foster a shared sense of historical experience. In Western Balkan capitals this is different than in Western European capitals. Balkan states have, like many other, but not all, CEE countries, a socialist history that they are not proud of. Czepczyński (2010) speaks about liminal (transformational) state: the landscapes are more typical for the previous regime and planning, but at the same times quite different from the aspired ones. 'The only and omnipotent communist parties ruled not only over economic, social, cultural life of a society, but also over the visualised and aesthetic expressions of everyday existence' (ibid., p. 69). The nation states that emerged from former Yugoslavia are young and have started recently with the process of (re-)constructing their national cultures. Bădescu (2016) states that capitals of former Yugoslavia were already regional capitals, had administrative and political roles before, but they were reshaped to sustain and strengthen their role as representative spaces for nation building in a politically unstable context and in a broader transition from socialism to capitalism. Former Yugoslav regional capitals did not have clear main squares; this concept did not fit in the communist planning theory and practice. Moreover, regional capitals were not supposed to have clear symbols of power—power belonged to the institutions of the capital of the country.

In their work on better understanding contemporary cultural and political change in post-socialist urban settings, Diener and Hagen (2013, cf. Young and Kaczmarek 2008) identify three themes around which the intersection of identity, nationalism and post-socialist urbanism has tended to crystallize and become manifest in urban places. These themes, which we adopt for our study of main squares, are the following. First, *selective remembrance and active forgetting*. There are broad tendencies toward crafting and commemorating revised and reified 'national' narratives by preserving, rebuilding or creating historical landmarks. These processes of reinterpretation often focus on projecting narratives of unity, tradition and perseverance back into pre-history. These new historical narratives are however highly selective and involve a significant amount of forgetting about alternative identity narratives. As mentioned above, this relates to the 'presentist' perspective of historical constructions that are constantly being reconstituted. The past is viewed the past with a (somewhat) limited to present-day set of attitudes and beliefs, the past can intentionally be interpreted in a certain manner or be forgotten. CEE countries have witnessed these processes of course during the socialist period (renaming of streets is a vivid example), but it is a universal process.

Second, *ambiguous spaces and banal practices*. Diener and Hagen mention that '… although proponents commonly deploy rhetoric that portrays efforts to build a "new" city or polity as straightforward, consensual, and all-inclusive, the inscription of new national narratives into the urban landscape is complicated by the presence of local, minority, and various "other" identities. These often manifest in divergent perceptions of urban iconography and spatiality. The presence of these alternative associations, practices, and memories problematizes efforts to construct homogenizing national narratives and results in ambiguous, contested, or contradictory urban spaces' (Diener and Hagen, p. 489). Especially in countries with ethnic and political divisions, the creation of a unifying space is difficult. In the Balkan, ethnic lines have been prominent in history. Diener and Hagen also suggest that '… socialist relics permeate the urban landscape, but their ideological foundations are largely forgotten amid the banal routines of everyday life and localized identities' (Diener and Hagen, p. 489). We referred to this point above—the once hated Casa Populori in Bucharest is now home of the Romanian parliament and 25 years after the transition, views of this place are changing and become more pragmatic. It is common that people forget the significance of for example statues and do not know who the person presented in the statue is.

Third, *globalized identities and cultural hybridities*. Post-socialist urbanism has featured efforts to reconcile traditional notions of local and national identity with new customs of regional and global integration. Cities were to become attractive places for foreign investment. Especially capitals should have an international appearance. And with modern media, international cooperation, foreign visitors and nationals going abroad, a degree of cultural hybridity developed. All over CEE, private business interests and international investments have driven urban redevelopment since the early 1990s.

Next to these three themes of Diener and Hagen (2013), in this chapter a fourth theme is discerned, namely the *intervention process*. Interventions of governments in countries with ethnic and political divisions can lead to protests—urban struggles even, because the creation of new symbols may come at a cost. This fourth theme informs about the importance of dynamics and the parties involved, the role of central government in local space, rules imposed, and the financial and immaterial costs involved. It is also about the possible support from society for new national symbols. A despised symbol—and CEE countries have a strong history and memory of such symbols—will perhaps last till a next government enters office.

Additionally, we consider to what extent the interventions in the square can be seen as innovations. In management literature, innovation is defined in various ways that generally come down to 'the process of creating value from ideas' (Tidd and Bessant 2014). Newness plus value are the two core elements of innovation. The concept of innovation is applied not only to business management, but to a wide array of themes, including landscape innovation and urban innovation. Incremental innovation is less risky than more radical forms of innovation. A difference between innovation in public space and innovation in business is that in public space 'value' is more difficult to identify and that innovations are more permanent.

12.4 Case Studies

Tirana's Central Square

Before 2003, Tirana's central square was a relatively simple transit open space,[1] and had never been a place for people to meet, it was rather a place for traffic regulation and authorized political meetings and/or protests (Aliaj et al. 2003). For understanding this space, it is useful to go briefly back into history. In 1912, Albania acquired independence from the Ottoman Empire. In search for a new identity, the new government of self-proclaimed King Zog prepared the first projects of European style urban design and architecture for the city centre, including a big new boulevard. The new square was symbolic and functioned as a conciliation space between the inherited organic city and the new modern Tirana. The neutrality of the square was much later further emphasized during the 1990s in an era of booming urbanization and a burst of informal urban development (Pojani 2014).

Pojani (2014) describes that since the 1920s up to date, successive governments have been trying to impose their own urban design visions towards a 'grand' city centre as a symbol of their power. This was always done top-down, through denial and erasing the built heritage of their predecessors, resulting in a main square that was never popular among all the people. During the communist decades, the main city square was considered 'official' and even dangerous by the people, and despite

[1] See Pojani (2014) for a detailed description of the history of Tirana's central square.

Fig. 12.1 Photo French Plan (photo: Besnik Aliaj, 2003)

some political events on the square, people avoided passing time on the square. During the post-socialist period, the authorities made several attempts to develop a contemporary 'grand' vision of the city centre and its main square, but efforts failed and resulted in political quarrels. The reasons can be found in the instability of state institutions and the rather chaotic nature of development of Albania's transition to a market economy. In 2003, an international design competition organized by the Municipality of Tirana resulted in a winning design (by 51N4E architecture studio, Belgium—see Fig. 12.1 and see Fig. 12.2 for the 2015 situation of the square), but it was only in 2017 before it was implemented. Indeed since 2003, several plans for the city centre have been made. The socialist party mayor of Tirana Edi Rama was in power between 2000 and 2011, and presented a masterplan for a new city centre, nicknamed the 'French Plan' by Architecture Studio Paris (2005). Before that, the city centre went through a process of cleansing informal constructions and activities that had emerged during 1998–2003, under mayor Edi Rama's 'clean and green programme'.

The French Plan included a sketch of a renewed Skanderbeg Square, surrounded by existing buildings and new high-rise towers, with a border of green and water splashes. The plan was refused by the then liberal-right wing government that was in power from 2005 till 2013. This government made its own plan during 2009–2013, reinforcing transit regulation, biking lanes and green. When Edi Rama, the mayor of Tirana, became prime minister in 2013, the plan for a pyramid-shaped square with a base of 160 × 160 m and a height at apex of 2.25 m (Fig. 12.2) was put on the agenda. The gravity of the square shifts from the Skanderbeg monument

Fig. 12.2 Skanderbeg Square, 2015 (Photo: Peter Nientied)

to the top (middle) of the geometric pyramid square. It was implemented in 2017, incorporating an underground parking, a central ring and a pedestrian space in the square. The interventions further included site clearance of informal kiosks, some parks and green spaces, as well as issuing permits for twelve high-rise towers and an eye catching central square. Our case study, the main square, is an element of a bigger redevelopment of the inner city.

How Did the Process Go?

The idea of prime minister Rama, who took up Tirana's main square as a personal project within his own governments programme 'Rilindja Urbane' (urban rebirth or renaissance), was to create an 'island' of monumental empty space in the chaotic and densified and formal/informal urban tissue of Tirana, symbolizing so the beginning of a 'new era' on planning and governance. The programme also wanted to show internationally that Albania is changing.

No bottom-up process with people or professionals took place. The planning system, legislation and implementation of the projects of 'Urban Rebirth' have been centralized. It is a national project, led directly by the prime minister office. Local government had no real say in the decision-making, at best it acted as an implementation arm of the central government. However, both supporters and opponents of the plans agree on the point that in a short period of time the government has been capable to implement the €12.5 million project.

Reinterpretation of the Skanderbeg Square, Selective Remembrance and Active Forgetting

The new square has transformed the heritage of the communist period, in fact wiped out the past. Pojani (2014, p. 88) asserts (in 2014, before the new plan was implemented) 'Like many centrally located public spaces, the centre of Tirana has served as a means of expressing political visions and ideologies. In the space of a century, it was transformed from an Ottoman market and prayer space to a scenographic colonial display, which included laic administrative buildings; to a monumental communist parade ground; to a post-communist anarchic marketplace; and finally, to a traffic roundabout surrounded by public office buildings, a bank, a large hotel, and a museum, but no human-scale spaces' (Pojani 2014, p. 88). Before the new intervention, the square was poorly treated, since 1991 the maintenance and new investments have been minimal.

Prime minister Edi Rama describes it in the following way during a speech in 2016: '… the square of Tirana is the brightest jewel in the crown of the National Program of "Urban Rebirth", which is actually bringing in Albania a transformation that is not just physical, but a cultural transformation. It is giving cities the opportunity to physically relate to the European identity of this country.'[2] The main argument brought to the public is ' … a city without a square, such as Tirana, is in fact a city without a community, a city whose identity has been hit right at its heart … Therefore, this justifies the start of project precisely with the logic to … democratize the vision which is there, and transform the space inherited, in a space where past, present and future coexist with each other … This means that … whoever becomes part of that space for an hour, a day or a lifetime, will get the sense of where they come from, where they are and where they are going …' (ibid.). This statement is interpreted as a signal towards an international positioning, in rather patriotic political terms.

A new main square, in fact an empty space, has been given an integrative symbolism. It has a focus on the future (notwithstanding the words of the prime minister on integrating past—present—future, 'history' on the square looks absent). The national symbol, the statue of national hero Skanderbeg (warrior on a horse as defender of Christianity and Europe) is not the focal point of the square any more, it became peripheral. Figure 12.3 shows the square when it was inaugurated—the statue of Skanderbeg is 'hidden' behind the podium. In the new square, the communist past is erased and existing buildings from that period (the museum and the opera house) have lost importance. The emptiness of the square and the topography of a low pyramidal shape oppose the monumentality of existing classic and modernist architecture (Fig. 12.4).

---

[2] https://kryeministria. al/en/newsroom/news/presentation-of-the-project-for-the-quot-skanderbeg-quot-square&page=41. Accessed 24 Dec 2017.

Fig. 12.3 Skanderbeg Square (Facebook Edi Rama, public realm)

New Narratives and Other Identities

The prime minister has given (cf. footnote 3) a detailed poetic narrative of the new square. It is about integration, linking past, present and future, the square as the heart of the community. The heart has been expressed in a big empty space for getting together. A question is to what extent the public will embrace this square as a heart of the community. In Albania, other identities do not play a big role, there are no ethnic minorities that are left out. A point at stake is however the lack of consensus regarding the square, and urban planning in general. The prime minister had a vision and secured support from his political majority at national and municipal levels. There has been little concern about the opinions of the political minorities—in this sense the square and urban plans are far from integrative.[3] Protestors point at legal and procedural mistakes, lack of involvement and the destruction of urban heritage, all for the sake of modernization—which is at the costs of democracy and respect for the rule of law. The process of Tirana's 'Urban Renaissance' project has been politicized, with—as usual—accusations about corruption and out of control finance. Like many CEE countries, and all Western Balkan countries, Albania seems to lack the capability to search for more consensus among different factions of politics and society.

[3] See for example the way in which a new Urban Plan 2013 was handled in the local council, http://www.balkaninsight.com/en/article/tirana-plans-the-transformation-in-a-metropolis-within-2030-12-28-2016.

Fig. 12.4 Skanderbeg Square (photo: Besnik Aliaj, 2018)

Ambiguous Spaces and Banal Practices

The new square is a multisensory experience; next to the visual experience and pyramidal topography (a sensitive symbol for Albanians because of the 'Pyramid', a mausoleum structure in honour of the past dictator), there are fountains with light arrangements, bikers playing, visitors making selfies, or just transit walkers passing. Despite activities of the municipality to provide animation (music, sports, etc.), the vast space of the square has remained mostly empty since its inauguration in June 2017. Indeed, temporary 'beer and kebab fairs' cannot consolidate a collective spirit of community on the square. The square is (not yet) a meeting place. Most young people prefer other more vivid spaces. During warm summers and windy winters, the empty square is not a very attractive space to spend some time. During a Muslim community celebration in 2017 on the new Skanderbeg square, the statue of Skanderbeg was hidden from the gathering, and the role and symbolism of Scanderbeg as a national hero and protector of Europe and Christianity was veiled. This led to critical comments.

Globalized Identities and Cultural Hybridities

The redeveloped square is a sign of a globalized identity. Foreign firms designed it, it's character is modern and western. One may even wonder how the redeveloped square is linked to Albanian identity—except for the stones that are supposed to have been collected from all over Albanian territory. From this angle, the development of place attachment of common people (the identification with the square as a national symbol and space) is not likely to develop in a short time. Given the volatile political developments in the Western Balkan, it is also quite predictable that a next government of a different political background is going to adjust the square.

Johannes Hahn, commissioner for EU enlargement negotiations, urged during a visit in November 2017 that EU wants clear proof from Albania regarding its ability to change. He said, during a visit to the Vlore Waterfront project (co-financed with EU funds) that such projects are the best 'visiting card' for Albania in Europe. The EU signal is for big 'modern' projects, not for cultivating national heritage and production on national resources. Tirana's new Skanderbeg square is such a visiting card. With the modernization of Skanderbeg square, Albania signals that its aim is to be a modern western nation.

Skanderbeg Square as Innovation

The redeveloped square is an innovation in the urban landscape. Its appearance—an empty space—is new, meant to be a symbol of Westernization and the idea behind it as integrative space is new. It is also a physical expression of a vision of the prime minister. While the appearance is new, the critical question is about its value—will the citizens of Tirana, of Albania, attach importance to this new space, will they develop a form of place attachment. Almost a year after its inauguration, there are few signs that the people do, but perhaps this process takes more time.

Conclusion

Tirana's main square project intends to enact a new image, a new identity of the city and the country. The new square was not a popular demand, but rather the perception of the prime minister, who has a personal, modern perspective of Tirana's space and its international orientation. Both the new narrative and the project of 'Urban Renaissance' have been disputed. The design, the corruption claims and the lack of participation hamper the identification of people with the new square.

Skopje's Macedonia Square

Skopje is capital of FYR of Macedonia—FYR referring to Former Yugoslav Republic. Before 2008, Skopje's central square was a rather simple open space where people met. In 1963 an earthquake hit Skopje and destroyed the city, and soon a new master plan was made under guidance of the Japanese architect and urban planner Kenzo Tange. The master plan projected a rapid expansion of the city (Stefanaovska and Koželj 2012), and had spacious designs with modernist architecture for public buildings (Deskova et al. 2017). After 1991, when Macedonia became an independent state (it was one of the six republics of Yugoslavia) Skopje grew rapidly—many people from the countryside moved to Macedonia's capital. Also, many Macedonians left the country. The diaspora sent remittances for investments in an apartment in the city, and, like in other capital cities discussed in this chapter, a lot of construction in housing and offices was created.

Since 2008, the city centre has changed dramatically. The nationalist-conservative Prime Minister Nikola Gruevski, who critics say ruled with an increasingly authoritarian bent during his decade in power between 2006 and 2016, presented plans for a new city centre, called Skopje 2014. This plan, 2014 referring to the year in which the implementation of the plan should be completed, was formally announced in 2010 and entailed the construction of some 40 monuments, sculptures, renovated facades and new buildings. This number has increased more than threefold during the process. Just on the newly constructed Bridge of Arts, there are 29 bronze-cast statues representing significant Macedonian male figures. The Skopje 2014 plan included the clearance of parks and green space, the construction of new government offices, buildings for public functions in the open spaces and an eye catching central square. Our case study is this square, as an element of a bigger redevelopment of the inner city.

The idea of the prime minister—who took up Skopje 2014 as his personal project—was to create a monumental inner city, that replaced modernist architecture with new neo-classicist buildings and new facades around existing buildings, new bridges crossing the river to extend the centre in the direction of the old town and bazaar and a huge number of monuments and statues honouring famous persons from Macedonia's history. Skopje 2014 has been well documented from the points of view of architecture, planning, project implementation (Mariotti 2016; Mattioli 2014) and the very high economic costs, and urban protests against the plan. The e-magazine Balkan *Insight* and *Skopje 2014 Uncovered*[4] gives details of the plans and the various projects. Both supporters and opponents of the plans and of the government agree on the point that in a short period of time the government has been capable to implement very far reaching interventions in Skopje's city centre, albeit at a high economic, social and political costs.

The interventions have transformed the heritage of the post 1963 period (see Fig. 12.5) in a dramatic manner. A central square has been created with new national heroes. The main statue that dominates the central square is 'Warrior on a horse', officially not referring to Alexander the Great. 'Everyone, including some tourist maps of the city, assumes 'the warrior on horseback' is actually Alexander the Great (see Fig. 12.6), although the conflict with Greece prohibits Macedonia from making this official' says The Atlantic's writer Armin Rosen (2013). And at a distance, on the other side of the river, a colossal monument has been devoted to the father of Alexander the Great, Phillip II, who symbolically looks at his son. The city square is a multisensory experience; next to the visual experience, there are fountains with

[4] www.balkaninsight.com and skopje2014.prizma.birn.eu.com/enresp.

Fig. 12.5 Central space Skopje in 2006 (source unknown)

light arrangements and there is constant music. On the main square there are also video-clips on the big screen of fashion stores and telecom providers.

Selective Remembering and Active Forgetting

The square was meant to be a symbol for a new national identity, which eradicated the recent socialist past. Macedonia's identity is a troubled one. Hopkins (2016) states in the *Foreign Policy*: 'Modern-day Bulgaria denies the existence of a separate Macedonian language; Serbia does not acknowledge the existence of a separate Macedonian Orthodox Church. And there is the country's ongoing bitter dispute with Greece, which continues to block the country's accession to international organizations like NATO and the European Union because of its name. (…). Macedonia was admitted only provisionally to the U.N. under the name "The Former Yugoslav Republic of Macedonia," and has altered both its flag and constitution in response to Greek pressure'.

In the process, the post 1963 period of modernist reconstruction of the city of Skopje is actively forgotten (wiped out is a better term) and on the ground it looks as if 40 years of Yugoslavian socialist period have not existed. Interestingly, after a detailed description of Skopje's history, the only bilingual Tourist Guide Skopje (2013) mentions only three facts that took place during the period (a large flood in 1962, the earthquake in 1963 and the new railway station in 1981). Nothing is said

Fig. 12.6 Statue warrior on a horse, Skopje main square (photo Peter Nientied, 2017)

about the socialist period, and nothing on the reconstruction after the earthquake. And much is said about the centuries before the socialist period. That is selective remembrance—the case is the creation of new narrative for the history, with Alexander the Great as founding father of Macedonia. Macedonia's national airport is Airport called Alexander the Great international airport. It is seen as a statement against Greece and fuels the political dispute on the origins of Macedonia. Skopje has a Museum of the Macedonian Struggle in its centre, and Greece has its own Museum of the Macedonian Struggle in Thessaloniki and claims that the name Macedonia already belongs to a northern Greek province, and that the name is a signal of the FYR of Macedonia's ambitions to claim territory beyond its borders. The point here is not which is the correct one—the point is that Macedonia created its own interpretation that should now be the leading narrative for FYR of Macedonia, and this narrative is expressed in the national square. The square's identity is a claim of the origins of Macedonia. That has given support among the people who are not fond of Greece's objections against FYR of Macedonia becoming an independent country with the name of Macedonia. In short, through the square the government aimed to rewrite history (see Fig. 12.7).

The plan Skopje 2014 goes with banal practices; some call it a kind of 'Disneyfication' of history. It attracts an increasing number of visitors, who find the square fun, impressive or surprising expression of history.

Fig. 12.7 Main square and bridge over the river (photo Peter Nientied, 2017)

New National Narratives and Other Identities

The new narrative is about the origins of Macedonia—that the foundations of Macedonia were in the present FYR of Macedonia—and not in the adjacent Greek province of Macedonia. Skopje 2014 was to give a 'correct' interpretation of history and tell the Macedonians what their true origins are. There is substantial evidence against this statement—the Slave origins of the Macedonians were even taught in all students' history books, but the government decided to make the Macedonian origins the key point of a new identity. The research of Kolozova et al. (2013) informs '[t]he expert focus group and interviewees are unisonous regarding the lack of aesthetics (labelled as "kitsch") and inadequate mythologization of history. According to the experts involved in this study, 'Skopje 2014' does not correspond with the predominant sense of national or ethnic identity and the perception of historical periods and aesthetics as formative of the Macedonian identity'. Based on a sample survey of 1400 respondents, the study of Kolozova et al. (2013) concludes that fewer than 10% of people saw Alexander as Macedonia's most important historical figure, nor did they feel an attachment to the era of antiquity he represents.

The new narrative is an ethno-nationalist view of Macedonian identity—which emphasizes a shared origin, language and culture. It is not about a multi-ethnic society; the Albanian and Turkish minorities are not part of the narrative. The Albanian minority party in the former government, Democratic Union for Integration (DUI), the largest Albanian political party before the elections of 2016, had an agreement that there would be three (3, out of 100+) statues of Albanian minority

Fig. 12.8 Hybrid culture (photo Peter Nientied, 2017)

members erected. It has not happened. Clearly, the Albanian and Turkish minorities are not part of the new Macedonian narrative. The Macedonian nationalist perspective denies that Macedonia is a multi-ethnic society, although 25% of the population is Albanian speaking, another 4% Turkish, 2.7% Roma and 4% others. Skopje 2014 has a detectably pro-family and even pro-natalist bent—almost all the woman it depicts are engaged in maternal activity. And there are only few statues of women. On the Bridge of Arts, 29 statues have been placed, all of them representing men.

National Identity and Globalized Identities

The neo-classicist spatial and physical expression of the new narratives goes with symbols of modern international society. On the main square, one finds a big modern building of a telecom company, and one finds a huge display for advertisements (see Fig. 12.8). The Warrior on a Horse watches, so to speak, these advertisements the whole day, and that is not considered as something strange.

Skopje 2014 was also meant to attract international visitors and this has worked out. As an example, Lonely Planet (2017) writes on its website: 'In the past few years, the central riverside area of Skopje (Скопје) has hammered out the look of a set design for an ancient civilization. Towering warrior statues gaze down on you and gleaming Italianate power buildings make visitors feel very small indeed. Marble-clad museums have mushroomed alongside hypnotic new mega-fountains. The building frenzy is the result of a controversial, nationalistic project called 'Skopje 2014'—still ongoing—implemented by the Macedonian government under Prime Minister Nikola Gruevski. Opinions aside, nobody can deny the Macedonian capital has become a thoroughly entertaining Balkan metropolis—a bit surreal and at times garish, maybe, but never dull'.[5] For a visitor the square is a lively square that is a kind of oddity in the city. Skopje may be called 'Europe's new capital of kitsch', but it attracts tourists and the square is a place to relax and meet, take photos in front of the fountains, etc. (cf. Graan 2013). It also appears that the younger generation does not care very much about the interventions—their concern is the bleak economic future of Macedonia and their hope is on finding an opportunity outside Macedonia, they are less concerned with the interpretation of their history.

Process

The planning and implementation of Skopje 2014 was centralized. Local government had no say in the decision-making. Laws have been changed and Skopje 2014 is a national project. The government has set on implementing its plan without any consultation. The Heinrich Böll Foundation (2016) shows in a video its critical take on the Skopje 2014 project, with due attention to the process. This source also explains the protests, from architects who saw their heritage destroyed, from opposition forces, from non-government organizations and from the citizens of Skopje who saw their city centre transformed. The struggle has been fierce at times (Véron 2016; various contributions on BIRN, cf. footnote 4). The process of Skopje 2014 was heavily politicized, with many accusations of corrupted practices. A secret service wiretapping scandal in 2015 started a process leading to the exit of the government and—after EU and US mediation—to elections end 2016. The government party remained the largest party, but opposition parties united and got a majority. It took more than half a year and again pressure from US and EU, before the majority could form a new government and the sitting government agreed to step back. The new government has started juridical investigations.

Innovation

Skopje 2014 can be seen as an urban innovation—not in the sense of modernization, rather as a so-called 'retro-innovation' cherishing history. The price of this innovation has been substantial, and whether this innovation leads to value creation that can justify the investments, can be doubted. What kind of public value has been created is a question too. Tourists like the new central space but among citizens it has led to rejection and disputes. Much money has been spent and the new government that assumed office in 2017 now thinks twice before demolishing the statues and the grotesque refurbishing of buildings. Discussions about distributing statues to municipalities and removing the huge Warrior on a Horse statue are now going on.

[5] www.lonelyplanet.com/macedonia/skopje/travel-tips-and-articles/skopje-behind-the-statues-a-beginners-guide-to-macedonias-capital/40625c8c-8a11-5710-a052-1479d276096e.

Conclusion

'The project "Skopje 2014" is not that much a project of preservation, or conservation, but rather the reconstruction of heritage and a peculiar invention of a heritage' says Janev (2014). Georgievska-Jakovleva suggests that 'Starting from the realization that there is no such thing as an inherent identity of a location and that material objects built on given location construe certain memories that then serve as spatial coordinates of identity, …, in the processes of interaction between place and experience, the "Skopje 2014" project should create a picture of the past that Macedonia would want to identify with, i.e., a picture that will be desirable to consume and shall ensure social cohesion, unified collective memory and, therefore, a stable identity' (Georgievska-Jakovleva 2014, p. 50). Through Skopje 2014, the state constructs an ethno-nationalist story of the eternal struggle of the Macedonian people for emancipation, thus 'proving' the legitimacy of the state (ibid.). Skopje 2014 is an attempt of the central political powers to formulate and communicate a new history of Macedonia and express a new identity. The central square and city centre were used to give an expression of this new identity, the built environment was used to display the new history. It is a history that is ethno-Macedonian and goes back to old times to create a new image of the origins of Macedonia. It has been hard for many Macedonians to relate to this new image. Both the new history and the plan Skopje 2014 have been disputed, for various reasons. Political opposition, the exclusion of minorities, the centralist management of the expensive project, scandals around the project and civic unrest (Véron 2016) are among the factors that impede the state from generating support and identification with this new identity. Skopje 2014 signifies a radical (brutal) intervention of the government, that cannot be easily turned back by next governments. To what extent the intervention will have an impact on the identification of Macedonian people with this new narrative remains to be seen. Meanwhile, the new government that took office in 2017 started a constructive dialogue with Greece, renamed the airport, and discusses the removal of the largest statue Warrior on a Horse.

Pristina's Central Square

After the 1999 Kosovo war, the withdrawal of Serbian armed forces and the establishment of United Nations Interim Administration Mission in Kosovo (UNMIK), Pristina became capital of Kosovo after the Republic of Kosovo declared independence from Serbia in 2008. Pristina has witnessed rapid urban development since 1999. The population increased because of an influx of people from other parts of Kosovo; Pristina had 205,000 inhabitants by 2012 according to official sources (Gollopeni 2016), but figures of 250,000 are also mentioned.

Like most of cities in the region, Pristina's city had a strong Ottoman identity. The bazaar (Çarshia) was the historic heart of old Pristina, the meeting place for people and the merchandising centre for locals and travellers since the fifteenth century. It was significant for Pristina's cultural and social development later. After WWII and with the new socialist regime, the development of the city centre and its function as the main city square shifted towards the western plain terrains. The city grew according to Yugoslavian modernist principles with wide streets, green public spaces and mass residential housing blocks. The modernization slogan 'Destroy the old, build the new' describes best the ideological climate at that time. The old bazaar was largely destroyed during the 1950s and 1960s. The city plan of 1953, according to its author Partonić (1953), did not foresee a main city square. Instead it promoted the development of a main avenue with pedestrian streets and main modernist administrative buildings along sides. In the city plan, the Yugoslav doctrine was visible, with a low urban profile and modest layout of public spaces, which were supposed to be the generators of community and public/social life, to keep things under control and keep Pristina as a simple regional centre. A big city square did not fit into this principle.

As replacement of the old city, a modest 'Brotherhood and Unity' socialist square was developed as part of the new modernist urban core, combined with a network of perpendicular streets and landmark architecture buildings. As Sadiki (2014) explains, the main street was considered as an avenue with bigger buildings along it sides, including the municipal headquarters, the Yugoslavian army headquarters, the regional parliament building, the post-telecom headquarters, the national theatre, Grand hotel, Sports palace, the cultural centre, national library, state energy company, etc. Only few historical buildings, like the mosque and the hammam, have remained. Although the socialist project of modernist Pristina and its city centre was never completed, it did create a new image and identity for the city as a reflection of socialist Yugoslavian values. The organic identity and Ottoman heritage of Pristina have been erased. The new political elite of Kosovo after its independence in 2008 was not interested to maintain the inherited modernist style of the city centre and its Yugoslavian square. There was discussion to remove the Brotherhood and Unity monument because it reminds people of the former enemy. The presence of international institutions and their point of not increasing tensions with Serbia hamper the wish to remove the monument.

After 2008, an uncontrolled process of privately initiated urban expansions was witnessed—the local public administration was weak and the construction sector strong (Boussauw 2012; Bejtullahu 2015). The term 'turbo architecture' and 'turbo-urbanism' (fast and informal urban infill development) have been used (Vöckler 2010). Pristina is in the 'disappearance act', says Hirt (2014, p. 125) about the change of public space to spatial privatization. With the Strategic Urban Plan of 2004, the Urban Development Plan 2013 and especially the economic crisis from 2008 onwards, the volume of uncontrolled development decreased—but the 'turbo architecture' is still going on, with limited attention for public space. Even the modernist green and open public spaces of the University of Pristina are gradually seized through informal occupation, fragmented restitution and privatization of land, and/or corrupted building permission. Another 'disappearance act' was the 'exodus' of an estimated 100,000 people in the period 2014–2015 from Kosovo to various European Union countries, in search of better prospects. And many Kosovars had already left the country before. In the first 6 month of 2015, almost 31,000 Kosovar (1.7% of the total population) citizens applied for asylum in Germany and Austria (König 2016). Only very few of them get asylum, and many 'disappeared' in the illegality in these two countries.

The city centre has a rather chaotic modern urban core without a clear identity. It has a network of roads, pedestrian pathways, fragmented green, and some landmarks from the communist and ottoman periods. Such broken collage of public spaces does not create a clear spatial order, and no clear cultural and political identity.

Pristina's main square is somewhat undefined (cf. Selmani 2012). Skanderbeg Square is in front of the major government building. On this square, the statue of Albanian hero Skanderbeg was placed in 2001 (Ermolin 2014). It is at the end of Mother Teresa Boulevard (see Fig. 12.11) that connects Skanderbeg Square (Fig. 12.9) to Zahir Pajaziti Square (Fig. 12.10).

These two squares and the Mother Teresa Boulevard (see Fig. 12.11) are the most important central public space. The space functions as main square of the country—commemorations and protests are performed. Political rallies are held in this space and so was the celebration of Kosovo's judoka Kelmendi who showed her gold medal to supporters during a welcoming ceremony in Pristina in 2016 and the first ever Pride Parade in Kosovo, called 'In the name of Love' in 2017, that started on Skanderbeg Square and ended on Zahir Pajaziti Square. The Boulevard and the two squares are first and foremost an area for strolling and coffees (cf. Norman 2014). It may be kept in mind that Pristina has benefitted economically from the presence of UNMIK—many foreign specialists have come to Pristina to help build institutions and keep peace. These foreigners have fuelled the market for housing, services and bars and restaurants.

The Process

At various occasions, changes have been made by the authorities (placing a statue, etc.), but most important was a redevelopment of the two squares and boulevard. After a period of plans and waiting, works on improvement of the central area of Pristina started in 2012, at an estimated cost of nine million euro—but higher figures have been mentioned too (Selmani 2012), with the idea to redevelop the squares and landmark buildings. The transformation in a pedestrian street, repair of war damage and painting facades, installing light and public booths and advertisements, etc. has not yet succeeded to give the area the functions of a popular city square. Also, the monument of Skanderbeg (which was welcomed by citizens) has not created a positive cultural and urban gravity that a main urban square can give to its own citizen. The improvements have been, in comparison to Tirana and Skopje, modest and rather functional. The public has not been involved in the planning and the professional community complained about the process (ibid.).

Fig. 12.9 Skanderbeg Square (photo Peter Nientied, 2017)

Fig. 12.10 Zahir Pajaziti Square (photo: Peter Nientied, 2017)

Fig. 12.11 Mother Teresa Boulevard (photo Peter Nientied, 2017)

Selective Remembrance and Active Forgetting

In terms of cultural landscape, Ermolin (2014, p. 169) concludes 'In less than twenty years Pristina has freed itself of almost the entire memory of socialist Yugoslavia and Serbian regimes. Instead, the capital of Kosovo has become full of visual tributes (by means of monuments and street names) to Skanderbeg, Mother Teresa, President Ibrahim Rugova, independence fighter Adem Jashari, US President Bill Clinton, and others' (Ermolin 2014, p. 169). These are symbolic acts of a new identity, a currently constructed Kosovan history. Indeed, the past period has been to do away with the physical expression of the Yugoslav and Serbian regimes. Street signs and inscriptions in cultural objects are still also in Serbian, but Demaj and Vandenbroucke (2016) show that 'Albanian ethnification' and the Albanian 'master narrative' dominate in everyday life. This has been very 'active forgetting'. Given the tense relationships with Serbia, the past is not forgotten, but it has been removed from the streets. In 2001 the statue of Skanderbeg was placed, and later other symbols of Kosovar identity followed. Given that Kosovo is a new state, a focus on rather ad-hoc national identity building is understandable. Hoxha et al. (2014) refer in their research to the national identity in planners' perceptions, and in their study, they confirm their hypothesis that 'if urban planners work during a nation-building phase, then their perceptions will reflect more elements of national identity' (ibid., p. 85). However, what this national identity is, is not clear as yet.

Ambiguous Spaces and Banal Practices, Other Identities

The central square in Pristina is unambiguous in the sense that it is an Albanian space and there is no room for influences of the Serb minority. Spatially it is, as explained, ambiguous. There is still discussion going on about the Monument of Brotherhood and Unity located just behind the Skanderbeg Square, whether it is a symbol of Serbian repression or a symbol of WWII. The mayor even proposed a referendum about the issue (Mari 2015), but that has not been organized.

Banal practices, as Diener and Hagen (2013) refer to, the ordinary things of life that take over a symbolic space, like drinking a coffee, are very much present. However, the point can be submitted that these banal practices are part of the symbolism of place and space. In other words, meeting and drinking a coffee together are elements of national identity. Pristina provides a central space for meeting and coffee and an evening stroll, and this is part of identity. 'As the ethnographic examples show, residents of Pristina set their mark on the cityscape through their small-scale everyday practices and uses of urban places' concludes Norman (2014, p. 8).

Globalized Identities and Cultural Hybridities

In Pristina, identities have been globalized. Since 1999 UNMIK has stationed many foreign staff and Kosovo has received help from many foreign organisations, who implemented all sorts of projects. Foreign influence has been substantial in the development of public organisations in street life. Second, Kosovo has already for a long time a substantial diaspora, not only bringing in monetary transfers, but also new mentalities and ideas, and foreign symbols. Names of target cities or countries of emigration are widespread, and in the centre the 'Swiss' figure a lot like Swiss hotel, Swiss casino, etc. (König 2016). Vöckler and Königgo one step further and, referring to the huge diaspora, wonder 'What constitutes the identity of a city that is the symbolic and institutional centre of a state that is increasingly transnational?' (Vöckler and König 2016, p. 21) The main squares and the Mother Theresa Boulevard reflect this. Its design and appearance are rather nondescript, at first sight it has no identity, an open space with some symbols, a space used by Pristina's citizens for meeting, strolling and coffee. But the spaces are used for commemoration of heroes, for political protest and for leisure. These functions, 'banal' practices, contribute to identity formation.

Innovation

The improvement of the central space of Pristina cannot be typified as an innovation, at most it is an incremental improvement of urban space. The old socialist identity has been removed, but a new identity has not been formed as yet. This is perhaps a sign of the troubled new state. The central space is not appreciated as newness by the citizens, but it does facilitate celebrations, protests, commemorations and daily practices of a stroll and a coffee. The space is intensely used, even though it is not special or new.

Conclusions

Today Pristina's centre is vivid and quite different from a decade ago, but its city core has still not been able to change its image as a peripheral centre of one of the six republics of former Yugoslavia. This is a complex question for the (divided) political elite. The interventions in Pristina's central spaces have been conducted without a plan of a main square as national symbol. This point is expected to come on the political agenda in the coming years, and then planners have the arduous task to find an expression of the dilemmas Kosovo currently faces: its orientation towards Europe and the EU and other global entities like Muslim League Countries, finding a story for a multi-ethnic and multi-religious population, the reality of extra-legality or establishing rule of law public institutions, and how 'Albanian' Kosovo consider itself to be.

Podgorica's Central Square

Montenegro's capital had in 2011 186,000 inhabitants in its capital city region and 151,000 in the city proper, according to the Statistical Office of Montenegro. To understand the process of transformation of Independence Square, a short explanation of Podgorica and Montenegro is useful (cf. Popović et al. 2016). Podgorica is the nowadays modern capital of Montenegro, after Cetinje—the historic capital. Between 1946 and 1992, during Tito's Yugoslavia times, the city was known as Titograd. The city expanded along the rich terrains of Zeta plain along Ribnica and Moraca river, which practically divide the old (Ottoman) and new (modernist) towns. The mixture of architectural and spatial identities in these urban parts reflects the turbulent history and identity of the city between different regimes, especially between Ottoman Empire (still visible in parts of Old town and Drac neighbourhood) and Yugoslavian Federation (new modernist town) (cf. Komatina et al. 2016).

Independence Square—formerly known as Republic Square, is part of the central town of Podgorica. The square is in New Town, which is the administrative, social and cultural heart of the city. The square covers of 1.5 ha, a relative modest surface compared to other main capital squares in the region (see Fig. 12.12, Independence Square is in the middle of the map). It is surrounded by the city library, the state art gallery, shops, etc. and has a rectangular shape (see Fig. 12.12). The square is bordered by lately renovated pedestrian zones and shopping streets. A pedestrian passage connects the Republic Square with Podgorica's City Hall and Montenegrin National Theatre building (Fig. 12.13).

Until 2006, the Independence Square was officially known under the name of *Ivan Milutinovic square*, a well-known WWII Montenegrin resistance leader con-

Fig. 12.12 Podgorica centre (photo: satellite image Google Satellites Pro, public realm)

sidered as national hero. By 2006, when Montenegro declared its own independence from what left of the Yugoslavian federation of Serbia and Montenegro, the square changed officially name as part of several political measures to establish the new national identity as an independent state, getting also out of international embargo consequences, because of wars in ex-Yugoslavia. Name change went through sensitive local and national political debates[6] but at the end it paved way for the square to undergo total reconstruction. It was widened, paved and a big central fountain was also built, while turning the area into a car-free zone. The project cost for reconstruction was about 2.5 million euros, including a beautification with colonnades, water lines and palm trees.

Selective Remembrance and Active Forgetting

After WWII, the process of reconstruction during communism led to the development of modernist massive residential blocks and social cultural and economic facilities (cf. Komatina et al. 2016). These residential and infrastructural developments shaped the layout of today's Podgorica and its spatial identity as well. The modest city square was part of that modernist Yugoslav identity, while the Ottoman heritage was almost erased. By the end of the 1990s, and especially after the declaration of independence in 2006, the city faced again a big transformation. New

[6] http://www.cdm.me/english/decision-made-republic-square-will-become-independence-square/. Accessed 24 Dec 2017.

Fig. 12.13 Independence square (photo: Peter Nientied, 2017)

residential buildings and modern business centres were developed. To create a new identity of a recognizable contemporary state and capital, the authorities made investments in the urban core's public spaces, including the main square. The city has renovated a network of squares, parks, pedestrian streets and monuments, where Independence Square is the most 'official' square, but not the most important and not the biggest public space. The square has no references to the Ottoman and communist identities. It aimed, as part of inner city redevelopment, to create space for a new identity as an independent European Montenegro, which historically has been a kind of Christian buffer region between East and West. The square is a space for public rallies and political meetings, but not the real meeting place for citizens. Meantime, today the new landmarks of the centre include new symbols: pedestrian and shopping streets, the orthodox cathedral, the Millennium and the Moscow bridges, several shopping malls, etc., dominating purposely the new Podgorica skyline, once dominated by the housing towers of communist grey modernist blocks. In the city centre renaming took place to erase memories of the communist past. It all started with renaming the capital, which after the war was renamed Titograd, after the Yugoslav leader, Josip Broz Tito. About 90% of citizens voted in a referendum in 1991 to return to the city's old name of Podgorica. The main street, Lenin Boulevard was also renamed Boulevard St Petar Cetinjski (a Montenegrin ruler and bishop from the eighteenth century) and many other streets have been renamed too. Renaming has become an issue in local politics between left and right, dividing also society and city debates to almost 50–50%.

Ambiguous Spaces and Banal Practices, Other Identities

Montenegro's capital main square received its fourth new name in a period of 70 years. It drew mixed responses from locals who had seen it all before. Many people have not been happy of these changes. Many consider these as practices related to the lack of institutional stability. Tomovic (2016) narrates: '… many people believe that renaming the city's main square means only one thing: a typical action taken when elections are clearly approaching. Unfortunately, these things bring votes, and not economic recovery or social projects … it merely stimulates also lots of unnecessary nationalist disputes'. While the ruling party in the city council justifies the move to rename the square by the need to reflect the reality of Montenegro's statehood, opposition parties and others accuse the authorities of using 'patriotism' to deepen divisions. Until WWII, the square was called *Union Square*, in honour of the unification of Serbia and Montenegro in 1918. After WWII, when communists took over Yugoslavia, it was renamed *Ivan Milutinovic Square*, after a communist political hero. On the eve of the referendum of independence on 2006, it was renamed a third time, as *Republic Square*. And early 2017 it was renamed into *Independence Square*. The names of other streets in the city were also changed at the same time. And after that, local architects with political links to government were hired to redesign streets, squares and public spaces. No public participation took place, and implementation of the projects was a typical top-down exercise. The style and atmosphere of the square is 'western'—the use of glass and steel facades mixed up with colonnades, water and palm trees. Even though the square often hosts cultural and entertainment events, it has been used during the last two decades symbolically for anti-government protests.

Globalized Identities and Cultural Hybridities

Because of its limited surface, the main square of Podgorica is less complex than the cases of Skopje and Tirana. Indeed, the space is more at a human scale; the 'cold' atmosphere of monumental concrete and asphalted materials in the past has been replaced by warmer and more colourful materials. The process of improvements and redevelopment of the main city square and the network of public spaces and pedestrian streets of Podgorica's urban core were aimed at giving the city a new European identity, to express that Montenegro is not a regional periphery of a former empire but should be regarded as an independent state. Although the main square transformation has been politically imposed top-down and without much respect for people's collective memory, the changes now are a fact and cannot be turned back. The square embraces a cultural and local-global hybridity: small scale and local, but with modern elements of capitalism and universal shapes and materials.

The Process

Independence square is part of a larger redevelopment project, as has been explained above. The modernization of the city centre was a non-contested large-scale intervention that was started just after the independence of Montenegro. As the square was not a major landmark, the process has been rather smooth. The renaming of the square led to more political fuss, it became part of the ever-present political quarrels.

Innovation

The city redevelopment is a successful innovation of urban central space. The main square is a rather unpretentious element of this overall redevelopment process. Podgorica has renovated its city centre and has created a modern, westernized core that is valued by the Podgorica citizens and visitors.

12.5 Conclusions

We have seen that the four main squares of capitals of small Western Balkan nations are part of a multiple and complex post-socialist transformation process (Table 12.1). It has become clear that main squares of capitals are significant symbols for nations. In the cases of Skopje and Tirana, very expressive symbols have been introduced. In the cases of the capitals of the young nations of Montenegro and Kosovo, the interventions have been less dramatic, but socialism was erased. In the countries' capitals that belonged to former Yugoslavia (Skopje, Podgorica and Pristina), main squares have not been part of their socialist history—a main square was not part of urban planning since it did not fit the planning ideology—regional centres were not to have main squares that could become a power symbol since power was with the communist party that had its power concentrated in the capital. In the case of Skopje for example, a square was created, before 2008 Skopje had a central space but no defined central square. All squares show globalised identities—as all four countries aim at EU membership and aim at business with new partners from the West—and investors from the Far East and the Arab world are welcome too of course. It was also evident that national governments, most visibly in the cases of Tirana and Skopje, oversee creating national symbols in municipal space.

The indicators of Diener and Hagen (2013) provided a useful framework for getting some order in the complex reality. In the following table, an overview is given of the four cases.

The main squares of Skopje and Tirana have been transformed and used as a place for expressing new national identities, with the ruling parties of the two countries in the lead—in both cases the prime minster has been the driving force behind the projects. A big difference between the two cases is that in Tirana, the new iden-

Table 12.1 Four main squares compared

	Tirana square	Skopje's square	Pristina square	Podgorica square
Reinterpretation: selective remembrance and active forgetting	Concept (identity and integration) less disputed than top down approach, design and implementation	Disputed nationalistic concept, wiping out history and creating a new history	Yugoslav/ Serbian past wiped out, Albanian history became lead story	Square as part of city centre modernised, westernized, past was erased
New national narratives ambiguous spaces and banal practices, other identities	New narrative of 'urban reconnaissance', 'all Albanian'. Banal practices on the square to make it come alive	New narrative. Other identities/ minorities not acknowledged—issue with Greece. Banal 'Disneyfication'	National Albanian narrative. Serb minority excluded Banal practices are part of identity	Part of larger modernization rather than a clear symbol
Globalized identities and cultural hybridities	'European' design and identity, international firms, national story and international symbol	Square done by international firms, gets international attention, is a tourism landmark	International orientation, modern ambiguous square	Modern and western appearance, without history
Process	Substantial funds, protests, limited involvement of local government and people	Huge funds, severe urban struggles, top down process, contract management shady	Unclarity on funds, municipal decision, no participation, no protests	Part of bigger redevelopment, after 2006 only renaming
Innovation	New square with yet unknown value	New square with disputed value	Improvement rather than innovation, intensely used	Part of urban innovation, valued by users

tity is modern and forward looking, and in Skopje it was a nationalistic narrative, going back centuries to find the 'authentic' Macedonian identity. In the case of Pristina, prominent has been the cleansing of Yugoslav and Serbian influences and the dominance of the Albanian narrative. In Podgorica, the square is a local rather than a national symbol and is part of the modernization of the city centre wiping out the communist heritage, in which it takes an otherwise rather unassuming position. Pristina and Podgorica are new capitals of new states that are very much in an early stage of state formation. The new states have serious national and international political issues to deal with both national between different ethnic groups and political parties and international. In Montenegro, the relationships with Serbia (and Russia) are a source of tension. In Pristina, Kosovo, the city's main square is somewhat nondescript, the delicate relation with Serbia is a motive for not erasing the last bit of the Yugoslav past, the monument of Brotherhood and Unity. Moreover, internally there is continuous political struggle between different political parties.

Possibly, the new state of Kosovo will give more identity to the main central space that we took as the main square, where the situation is more settled, and a clear political majority is in power.

The cases of Tirana and especially Skopje are quite different. The governments of Macedonia and Albania have redeveloped their main squares as symbols of national identity according to political party elites in a globalized world. In 2016, the national government in Macedonia was in power for a number of years and had the political majority and time to develop the main square (city centre) as a new symbol of Macedonia. Difficult relationships with neighbouring countries did not prevent the then government to spend a lot of energy and money on the new square. It has become a major transformation, which is a 'headache dossier' for the government that came into power in 2017; what could it do with this heritage?—destruction is costly. In Albania, the socialist government was in power in 2017 for 4 years and just before the elections, the new main square was inaugurated and many popular activities were organised to attract crowds to the square. The socialist government was re-elected and now also has a majority in the Tirana's council. This means that for the coming years no new intervention will take place in the square and that the Albanian people have some years to attach themselves to the new square. Whether this will happen remains to be seen.

In short, all four countries show dynamic political processes, with limited respect for the past and at times fierce struggles. In these conditions, a question is what a next government will do with the square developed by its predecessor. This is an actual question in Skopje, but in Tirana it may take some time before the opposition is back in power. In Montenegro and Kosovo, the squares are not examples of visions of prime ministers and are unlikely to be modified.

In none of the cases, much attention has been paid to place attachment of citizens. Design principles to help people develop place attachment (Lewicka 2011; Evans et al. 2011; Bridge and Watson 2013) have not been applied. In all cases, the design and redevelopment of the square was based in a top-down fashion on how a square should look like and the narrative it should express, rather than questioning what would help people to develop place attachment and embracing the new development as a part of their identity. In the Podgorica case, the improvement of the square was part of a wider modernization of central urban space and this modernization (away from communist memory) was the new national notion. In the case of Skopje, the plan Skopje 2014 was an icon of the nationalist government and Skopje 2014 was not merely a spatial redevelopment, but a very political intervention. In the Tirana case, there was political resistance against the socialist government of Edi Rama (in 2017, just before the inauguration of the square) and the square became part of the overall political tensions.

The final point of this chapter concerns the framework of Diener and Hagen (2013), who edited a volume of papers that signify a major contribution to the exploration of relationships between urban space and national identity. We have added a fourth component 'how did the process go' to the framework, because it was thought to be a meaningful factor. In our study, it was. Another issue regarding the framework is the treatment of 'banal practices', which have a certain ambiguousness. Banal practices can be taken as everyday activities (commonplace culture, the stroll and coffees in Pristina), or as the 'kebab & beer fairs' in Tirana and the 'Disneyfication' of space in the Skopje case. It makes sense to distinguish these two sides of 'banal' practices. In the Balkan context, the word 'banal' has a somewhat negative connotation. Everyday activities however help to foster place attachment, and thereby embracing an adjusted identity.

References

Aliaj B, Lulo K, Myftiu G (2003) Tirana, the challenge of urban development. Co-Plan/SEDA, Tirana

Bădescu G (2016) Capital city-making: a bird's eye view of independent urbanism in former Yugoslavia and the Soviet Union. MONU Mag Urban 25:8–15

Begić S, Mraović B (2014) Forsaken monuments and social change: the function of socialist monuments in the Post-Yugoslav space. In: Moeschberger SL, DeZalia RAP (eds) Symbols that bind, symbols that divide. Peace Psychology Book Series. Springer, Berlin, pp 13–37

Bejtullahu FH (2015) Demand for housing quality and urban liveability, potential for establishing a new identity of city (Prishtina). University of Pécs, Department of Breuer Marcell Doctoral School of Architecture (Unpublished paper)

Belange D, Casaló LV, Flavián C (2017) Understanding the cognitive, affective and evaluative components of social urban identity: determinants, measurement, and practical consequences. J Environ Psychol 50:138–153

Bollens SA (2009) Intervening in politically turbulent cities: spaces, buildings, and boundaries. J Urban Technol 16(2-3):79–107

Boussauw K (2012) Challenges, threats and opportunities in post-conflict urban development in Kosovo. Habitat Int 36(1):143–151

Bridge G, Watson S (2013) Reflections on affect. In: Bridge G, Watson S (eds) The new Blackwell companion to the city. Wiley Blackwell, Chichester, pp 277–287

Cheshmehzangi A (2011) Identity and public realm. Procedia Soc Behav Sci 50:307–317

Cheshmehzangi A (2015) Urban identity as a global phenomenon: hybridity and contextualization of urban identities in the social environment. J Hum Behav Soc Environ 25(5):391–406

Czepczyński M (2010) Interpreting post-socialist icons: from pride to hate towards disappearance and/or assimilation. Hum Geogr 4(1):67–78

Demaj U, Vandenbroucke M (2016) Post-war Kosovo landscapes in Pristina: discrepancies betweenlanguage policy and urban reality. Natl Pap 44(5):804–825

Deskova AO, Deskov V, Ivanovsklі J (2017) Inclusiveness as urban and architectural concept, the case of the postearthquake renewal of Skopje. In: Marina O, Armando A (eds) Inclusive/exclusive cities. Politecnico di Torino, Turin, pp 172–191

Diener A, Hagen J (2013) From socialist to post-socialist cities: narrating the nation through urban space. Natl Pap 41(4):487–514

Ermolin DE (2014) When Skanderbeg meets Clinton: cultural landscape and commemorative strategies in postwar Kosovo. Croat Polit Sci Rev 51(4):157–173

Evans B, McDonald F, Rudlin D (eds) (2011) Urban identity. Routledge, Albingdon

Ferenčuhová S (2012) Cities and urban research in postsocialist Europe. In: Enderdon T, Jayne M (eds) Urban theory beyond the west, a world of cities. Routledge, Abingdon, pp 65–74

Georgievska-Jakovleva L (2014) The project "Skopje 2014": between social cohesion and social disintegration. Culture 8:43–55

Gollopeni B (2016) Socio-urban developments in Kosovo: Study case Pristina. Int Sci J 6:81–93

Graan A (2013) Counterfeiting the nation? Skopje 2014 and the politics of nation branding in Macedonia. Cult Anthropol 28(1):161–179

Heinrich Böll Foundation (2016) How the Skopje 2014 project ate the urban commons. www.boell.de/en/2016/11/07/how-skopje-2014-ate-urban-commons. Accessed 2 April 2017

Hirt S (2014) The Post-public city: experiences from post-socialist Europe. Refereed Proceedings of the 102nd Annual Conference of the Association of Collegiate Schools of Architecture - Globalizing Architecture: Flows and Disruptions, pp 123–129

Hopkins V (2016) Let them all eat Alexander the Great statues. Foreign Policy. http://foreignpolicy.com/2016/06/19/let-them-eat-alexander-the-great-statues-skopje-2014-macedonia-colorful-revolution/. Accessed 12 April 2017

Hoxha V, Dimitrovska Andrews K, Temeljotov A (2014) Cultural factors affecting urban planners' intentions to regulate public space in Prishtina, Kosovo. UrbaniIzziv 25(2):76–89

Janev G (2014) Skopje 2014: instrumentalizing heritage for unexpected results. Friedrich Schiller University Jena, Cultures of History Forum, ImreKertészKolleg. http://www.cultures-of-history.uni-jena.de/debates/macedonia/skopje-2014-instrumentalizing-heritage-for-unexpected-results/. Accessed 12 April 2017

Kolozova K, Lecevska K, Borovska V, Blazeva A (2013) Skopje 2014 project and its effects on the perception of Macedonian identity among the citizens of Skopje. Institute of Social Sciences and Humanities, Skopje

Komatina D, Kosanović S, Aleksić J (2016) Urban devastation: the case study of Podgorica, the capital of Montenegro. Archit Urban Plann 12(1):52–58

König J (2016) Prishtina: Departure City? Eurozine - www.eurozine.com/pristina-departure-city/?pdf. Accessed 15 Aug 2017

Lewicka M (2011) Place attachment: how far have we come in the last 40 years? J Environ Psychol 31:207–230

Light D, Young C (2013) Urban space, political identity and the unwanted legacies of state socialism: Bucharest's problematic Centre Civic in the post-socialist era. Natl Pap 41(4):515–535

Light D, Young C (2015) Local and counter-memories of socialism in post-socialist Romania. In: Beyen M, Deseure B (eds) Local memories in a nationalizing and globalizing world. Basingstoke, Palgrave, pp 221–243

Lisiak A (2009) Disposable and usable pasts in Central European Cities. Culture Unbound 2009:431–452

Lonely Planet (2017) Introducing Skopje. Retrieved from http://www.lonelyplanet.com/macedonia/skopje/introduction. Accessed 14 April 2017

Mari C (2015) A monument between brotherhood and unity and Adem Jashari. Kosovo 2.0, October 30, 2015. http://archive.kosovotwopointzero.com/en/article/1926/a-monument-between-brotherhood-and-unity-and-adem-jashari. Accessed 15 Aug 2017

Mariotti J (2016) What ever happened to Skopje? Monu Mag Urb 25:4–7

Mariotti J, Koželj J (2016) Tracing post-communist urban restructuring: changing centralities in Central and Eastern European capitals. Urbaniizziv 27(1):113–122

Mattioli F (2014) Unchanging boundaries: the reconstruction of Skopje and the politics of heritage. Int J Herit Stud 20(6):599–615

Murzyn-Kupisz M, Gwosdz K (2011) The changing identity of the Central European city: the case of Katowice. J Hist Geogr 37(2):113–126

Norman K (2014) Shifting experiences of places in Prishtina. Eurozine. www.eurozine.com/shifting-experiences-of-places-in-prishtina. Accessed 15 Aug 2017

Osborne BS (2001) Landscapes, memory, monuments, and commemoration: putting identity in its place. Queen's University, Department of Canadian Heritage, Kingston

Partonić D (1953) GeneralniUrbanistički Plan Prištine –General Urban Plan of Pristina. Municipality of Pristina, Pristina

Paskaleva K, Cooper I (2017) Forming post-socialist urban identities through small-scale heritage-based regeneration: a role for intangibles? J Urb Des 22(5):670–688

Pojani D (2014) Urban design, ideology, and power: use of the central square in Tirana during one century of political transformations. Plann Perspect 30(1):67–94

Pojani D, Maci G (2015) The detriments and benefits of the fall of planning: the evolution of public space in a Balkan post-socialist capital. J Urb Des 20(2):251–272

Popović S, Lipovac N, Vlahović S (2016) Planning and creating place identity for Podgorica as observed through historic urban planning. Prostor 24(1):63–73

Rosen A (2013) Skopje's strange quest to rewrite its history ... Through public sculpture. Citylab. http://www.citylab.com/design/2013/08/skopjes-strange-quest-rewrite-its-history-through-public-sculpture/6416/. Accessed 21 Nov 2017

Sadiki A (2014) "Značajdruštveno-kulturološkograzvoja u arhitekturiPrištine u razdoblju od 1945 do 1990. godine" (The influence of the social-cultural developments in the architecture of Pristina during 1945-1990). University of Belgrade, unpublished PhD paper

Selmani A (2012) Critics say revamp takes Pristina back to square one. Balkan Insight. www.balkaninsight.com/en/article/critics-say-revamp-takes-pristina-back-to-square-one. Accessed 15 April 2017

Sepe M, Pitt M (2014) The characters of place in urban design. Urb Des Int 19(3):215–227

Stefanaovska J, Koželj J (2012) Urban planning and transitional development issues: the case of Skopje, Macedonia. Urbaniizziv 23(1):91–100

Sýkora L, Bouzarovski S (2012) Multiple transformations: conceptualising the post-communist urban transition. Urb Stud 49(1):43–60

Ter-Ghazaryan DK (2013) "Civilizing the city center": symbolic spaces and narratives of the nation in Yerevan's post-Soviet landscape. Natl Pap 41(4):570–589

Tidd J, Bessant J (2014) Strategic innovation management. Wiley, Chichester

Tomovic D (2016) Montenegro square in line for fourth name. Balkan Insight, 7-3-2017. http://www.balkaninsight.com/en/article/montenegro-square-in-line-for-fourth-name-03-04-2016. Accessed 24 Dec 2017

Tourist Guide Skopje (2013) Trimax Publisher, Skopje

Tuvikene T (2016) Strategies for comparative urbanism: post-socialism as a re-territorialized concept. Int J Urb Reg Res 40(1):132–146

Véron O (2016) Contesting the divided city: arts of resistance in Skopje. Antipode 48(4):1441–1461

Vöckler K (2010) Turbo-urbanism in Prishtina. In: Dmitrieva M, Kliems A (eds) The Post-Socialist City: continuity and change in urban space and imagery. Jovis Verlag, Berlin, pp 210–229

Vöckler K, König J (2016) A translocal capital? Prishtina. Monu Maga Urb 25:16–21

Young C, Kaczmarek S (2008) The socialist past and postsocialist urban identity in Central and Eastern Europe, the case of Łódź. Eur Urb Reg Stud 15(1):53–70

Chapter 13
Professional Associations as Public Actors in the Formulation and Implementation of Spatial Development Policies. Key Study Monitoring and Evaluation Standard for the Urban Environment of Sofia

Angel Burov, Vasil Madzihirski, and Irina Mutafchiiska

13.1 Introduction

During the transition period in Central and Eastern Europe, large power imbalances and asymmetries on national and local level were retained on Bulgaria's development path (REC 1998). Even though the country joined the EU in 2007 with high expectations towards democratization, there are still great challenges related to the division of power (EC 2017), decentralization and local community empowerment. Recent studies provide evidence about the "capture of the state" (CSD 2016; IME 2017) with major aspects being outlined such as the dependence of the judicial system and the spreading of the corruption. These two aspects have had profound political impact over many recent environmental cases both out or in urban environments (Let Nature Remain in Bulgaria 2018a, b) and to the access to justice on environmental matters (UNECE 2018).

The role of the non-political actors is described as crucial for the challenges in front of the democracy in the country (IME 2017). Along with the non-governmental organizations, civic movements and the business community, the professional associations are included as potential actors in the treatment of important societal affairs. Segregation and land markets are such affairs that have long-term impact upon the urban communities and their living space. A related policy paper prepared for and discussed at the United Nations Conference on Housing and Sustainable Urban Development Habitat III (UN 2016) stresses upon the special role of the national-level professional associations focusing on urban and territorial development in the policy design, implementation and monitoring of sound urban spatial strategies. Some of the key recommendation about the monitoring couple it with adequate

A. Burov (✉) · V. Madzihirski · I. Mutafchiiska
Association of Bulgarian Urban and Regional Planners (ABURP), Sofia, Bulgaria
e-mail: angel.burov@bgplanning.org

© Springer International Publishing AG, part of Springer Nature 2019

M. Finka et al. (eds.), *The Role of Public Sector in Local Economic and Territorial Development*, EAI/Springer Innovations in Communication and Computing, https://doi.org/10.1007/978-3-319-93575-1_13

participative procedures needed in contexts where speculation, gentrification and displacement are taking place.

During the transition period and especially at the end of the 1990s, the public sector became more reformative in the state and local economic matters but the territorial development planning has seen almost full abandonment in the early 1990s as something unpopular from the socialist past and it was not seen as instrument for restructuring and reshaping the country and its settlements. Its resurrection came partially in the late 1990s and in the beginning of the New Millennium slightly different from the way spatial development policies were formulated and implemented a decade ago. Except for the inviolability of the private property guaranteed in the Constitution of the Bulgarian state, the planning system didn't see many innovative solutions and its full array of legal instruments is still under preparation and not much integrative as a whole. In the last two decades, the local authorities started to receive new competences and responsibilities without well-designed financial mechanisms and one of the means to cope with the situation was the privatization and sale of key portions of public land and assets.

There has been general deficit of thorough applied research and expertise in issues related to land rights and prices and access to infrastructure, distribution of population and spaces for appropriate functions and services, urban decay and brownfields redevelopment, ageing of the built-up fabric and its life cycle, environmental pressures and external costs, etc. This has led to the poor quality of the public policies for spatial development and to enormous benefits for narrow private interests that took advantage.

The fast growth, the long political-economic transition, the scarce budget and funding for Bulgaria's capital Sofia and its urban environment along with the uneven development landscape at different speeds have led to great contrasts in the city. Today many neighbourhoods exist in a self-reliance state for decades while other areas experience consecutive transformations and private investment pressure rarely leading to better quality of the public space. Combined market and state failures aggravate the living conditions of many old and new residents in the most attractive and most neglected urban areas. This long-held urban crisis has mobilized many actors pursuing individual and common interests.

The Association of Bulgarian Urban and Regional Planners (ABURP) is one of the new actors, getting involved in the debate for the spatial development of Sofia Municipality and its metropolitan area. It is trying to demonstrate the benefits from a more open process with broader stakeholder consultation, involvement and networking along with the expertise needed in the elaboration of spatial development policies. Some of the values of the organization stated in its statute are focused on improvement of the professional and public environment for the spatial and urban planning activities, around which informed and reasonable decisions are taken and planning culture is being built.

The key study in this text looks at two recent demonstration projects of the ABURP which raise many questions for the role of the professional associations and their impact in society, the strategic planning in action and the models for shared development, the urban knowledge and standardization of monitoring and evaluation. These specific questions are considered as part of more broad issues like the institutionalization of planning, the communication and mediation in the governance process, the power and rationalization in spatial development policy formulation and implementation.

13.2 Theoretical Overview: The Planning Profession, Its Institutions and Planning in Action

The improved knowledge is an objective (EU 2016) at the base of the deeper understanding of spatial development processes where urban and regional planners and similar professionals can help in building it. Such knowledge base can produce more consistent urban knowledge arenas (COST Action C20 2009) with debates that go far above the simplistic and populistic nature of the usual public hearings hardly climbing up to the first rungs of tokenism along the ladder of citizen participation (Arnstein 1969).

We hold that the constant communication and participation and in-depth monitoring and evaluation are the basis not only for better knowledge but also for wiser decisions and suitable solutions. These are activities that demand active and socially balanced involvement into peaceful struggles through research for evidence and proof of concepts. And this could be true only if decision makers, experts and the interested public are ready for debates owing to multiple and complexly linked preconditions. These preconditions are out of the scope of this chapter but for instance well-articulated in the classic and seminal essay of Patsy Healey about planning through debate (Healey 1992) with its challenging propositions.

Spatial development and planning are processes much influenced by political-economic and political-ecologic stances and the power concentrated along decision-making networks in the societies (Flyvbjerg 2002; Allmendinger 2002; Hacking and Flynn 2017). The urbanization of capital takes shape that is rarely in the hands of the citizens in liberal democratic states with dominating capitalist mode of production and consumption (Harvey 1985). The formulation and implementation of spatial development policies ranges from research, through planning, public incentives and regulations to public and private investment dedicated to the development.

Along this spiral many strands of expertise are necessary. The one that settles the organization of the diverse disciplines of thought and action and legitimizes the government decisions is the planning profession which is at the heart of the development process. Those who practise planning as professionals possess certain skills, experience and knowledge in regard to both the process and the content in the specific context.

Beth Dempster (1998) gives a multi-layered definition of planning:

"The term planning can be applied to an activity, a process, a profession, and/or a discipline. As an activity, its primary concern is to visualize future possibilities and intentionally choose, guide, and/or create current behaviours, structures, and/or tools to achieve and/or target toward desirable future states. As a process, it refers to the methods, mechanisms and tools for accomplishing the activity, especially in a group or social context. As a profession, it refers to the group of individuals carrying appropriate skill sets who fulfil an agreed upon

social responsibility to guide the processes. As a discipline, it refers to the body of knowledge, related to all aspects of planning, held symbolically or in the minds of practitioners, researches, and theorists".

As Philip Allmendinger (2002) states about the Anglo-Saxon context:

"Once planning had a professional institute distinct from the engineering and architectural professions it followed the same path as other professions in establishing a symbiotic relation between itself, its members and the government".

Which he continues that is the basis of reciprocal relationship:

"The government needs planners to undertake its policies. Planners and their professional body need the government as an employer and to legitimize their claim to professionalism (and associated status and benefits)".

We argue that no matter the origin of the profession and its distinction in the different contexts the government can find the professionals that suit its purposes but also its decisions can be influenced by various professional groups to a certain extent. In times of changes, transitions or critical social conditions, there is risk for the politically powerful to lose parts of their legitimacy. They can retain power either through fearful manipulation or through creative adaptation. Such time provides good chances for moving up along the ladders and spirals of political evolution if citizens are stubborn enough to overcome the manipulation and the professionals to actively promote alternatives. This is also a good time when autopoietic expert systems such as the spatial planning system can turn into sympoietic systems in the meaning attributed to the term about planning (Dempster 1998). Sympoietic systems provide space for various sources of expert and experiential knowledge that can bridge both the theory-practice epistemological gaps but also the substantive and procedural gaps in the planning and spatial development strategies and actions. This is really important when established models prove to be inadequate to the dynamic changes and the information arrays insufficient for the new conditions.

The idea of planning in action is present for a long time in the theoretical and practical discourse of the planning profession. John Friedman's classical view for transactive planning from 1973 is corresponding to other reformation works of the rationalist planning framework from that period and it has been reinterpreted and further explained in several consecutive works of the author (Friedmann 1973, 1987, 2003). Many others have added new perspectives and brought ideas from other fields of thought in this direction calling for mutual understanding of interests and dialogue through participatory planning in the face of power (Forester 2007) communicative and collaborative planning (Healey 1997). Close to the idea of the transactive planning theory but taking place in the theoretical field of economics is the idea of transaction costs. It was raised up to a body of theory known as the transaction-cost economics (Williamson 1979). Through this idea we come to the issue about joint agreements or even public consent related to the (re)use of land and property, (re)organization of territory and jurisdictions. In the organization of such decision-making processes, there is great need for communication and collaboration considering the amount of property owners, inhabitants, businesses and users of private or public services and assets involved in the spatial development process. In order to transform multi-faceted views about spatial development, knowledge about spatial phenomenon and attitudes towards the (re)use of space into real action common denominators have to be found. This is even more necessary if societies and communities take seriously into account the good governance approach for democratic choice, the equity and efficiency principles in socio-economic organization, the environmental justice and long-term sustainability of development.

13.3 Conceptual Context: The Transactive Concept in Spatial Development and Lower Level Institutional Thickness

Considering the theoretical overview we take a glimpse at the vertical and horizontal axes of political action along which two parallel concepts have taken shape—the multilevel governance and the institutional thickness.

The first comes from the European integration view of subsidiary political entities that formed in the context of national hierarchically structured administrative-territorial units (Piattoni 2009). It is also used in the context of overlapping competences of institutions working around multiscale socio-ecological systems nested vertically in a complex panarchy (Gunderson and Holling 2003). There is often uneven distribution of power along the levels or at the different scales including sovereign and autonomous entities as well as joint and mixed groups of political bodies with different jurisdictions, mandates and responsibilities.

The second concept comes from the view of horizontal proximity or density of groups of actors (Amin and Thrift 1995), structured more or less formally in various representative or other institutions. These organizations are making common decisions given their distributed power or self-empowerment along with the property rights, tangible assets and other resources they manage or rule.

Both of these ideas help us analyse the gaps related to the transmission and transposal of strategies into actions when formulating and implementing spatial development policies aiming at the reproduction or new production of decent living conditions in the urban environment and human habitats. The key study illustrates such analysis and the synthesis that gaps could be overcome by investing in human resources and institutional building on lower levels along with relevant property rights upon the common goods (Alexander 1977).

Yet, coordination and communication mechanisms and integrated policies application, monitoring and evaluation involving the lowest and intermediate levels well connected up to the metropolitan scale are important. This could enable interested parties: (a) to make cross-comparison of the dynamics in the development state of different places; (b) to investigate the internalized and externalized impacts that are positive or negative in and around the cities; (c) to correlate the satisfaction levels among the citizens and the conditions in their areas. The outcomes expected from such approach could be more interest in participation or better representation, eventual reaching of consensus or co-creating of local community and networking initiatives, combining both hard and soft measures with the eventual support of metropolitan and national urban policies. The impacts of such initiatives can help to build capacity for self-reliance in the neo-liberal context and more general resilience (Davoudi et al. 2012) to the development contradictions (Campbell 2003) at stake. These contradictions are either place specific for a local community's space or come from globally flowing accumulated capital that leads to creative destruction which affects certain types of spaces in attractive or suitable to exploit locations on their way (Castells 2008). The promising solutions to the contradictions should be more tailor-made than one size-fits-all. The metropolitan governance system has to be able to look before and after market and state failures. It also has to be able to overcome planning and design pitfalls leading to urban decay including unhealthy living conditions, unsafe urban environment, unsatisfactory arranged or shaped, equipped or maintained built-up areas, non-liveable neighbourhoods and poor ecology of the street life (Appleyard et al. 1981), as well as ruined or substituted traits of past and present creativity.

Thus the costs for filling the gaps have to be considered in advance and taken into account in every unique spatial planning system—national, federal, regional. Both during its reforms and in the reorganization of local administrative structures, the introduction of co-design or co-implementation models for the urban communities and neighbourhoods is a must. Some of the major costs for this are related to certain people's time with their knowledge and skills to use appropriate tools. Such resources could be dedicated to the acquisition of information, provision of communication, advocating or resisting, design thinking, urban planning and design, fund raising and/or application for funding, etc. Certainly, the links between the people involved on the lowest levels and the public authorities remain important question both as job descriptions, contractual relationships, legal statutes, funding mechanisms and technical equipment. And it is up to the specific cultural contexts of interplay and openness which calls for a wide spectrum of solutions that could be outlined, debated and introduced.

All of this constructs the conceptual framework behind the two recent and consecutive demonstration projects of the ABURP realized in Sofia, Bulgaria. Both projects were dedicated to the existing urban environment—the promotion of its liveability and values and the monitoring and evaluation of its quality through socio-spatial research and benchmarking.

13.4 Key Study: ABURP's Demonstration Projects "Share the Neighbourhood" and "Set an Urban Standard" in Sofia

The Spatial Development Policies and the Planning Context in Bulgaria and Sofia

In post-communist countries, old models in the realm of spatial planning still exist and they can be found in planning culture, administrative practices, legislation and in social attitudes. Fundamental characteristic of the communist period was the leading role of the state authority in the planning process. The power was centralized in the hands of few people and they practically took decisions about every aspect of public life, including the profile and location of industries, agriculture, urban settlement network development, housing policy, social infrastructure, etc. Architects often specializing in urbanism had the leading role in projecting and shaping socialist policy about spatial planning, land use and urban design. The planning approach and instruments were following modernist technocratic principles although distorted by the specific social stratification regime of Soviet and Eastern European countries at that time (Szelenyi 1983). Strict bureaucratic rules in territorial planning and land use practices were established but exemptions were allowed under the privileged position of highest level expert commission which was indeed sanctioning disputable cases in a political manner.

One of the biggest challenges in the post-communist transition period was to act with previously established instruments and rules in new conditions. The statute of larger scale planning was severely questioned, private property was reinstated and new values appeared. One of the newly settled principles was that the new owners have the right to do whatever they want with their property in an extreme liberal fashion. The will of the owners quickly took precedence over planning instruments used mostly for partial amendments. Although outdated as approach these instruments were being kept as means of bureaucratic control and blackmail, additionally providing quasimarket conditions for urban design services for architectural and judicial consultants that are currently tight with the administrative establishment. Restitution was allowed through the "real borders" of historical allotments, many of which were previously agricultural land and now in the outskirts of settlements. Controversial privatization took place and very small proportion of reserved public land stayed in the portfolio of state and local authorities. This was accompanied with decentralization process after which local authorities received broad political and administrative autonomy but few financial and taxation instruments in their hands. They didn't have any experience and resources in self-governance, including such for the formulation and implementation of spatial development policy. Thus quarter century later the results of these circumstances can be clearly seen in the patchy urban landscape of Bulgarian cities. Sofia which has been the most dynamic city during the transition period experienced fully all of its disadvantages affecting plenty of old and new residents.

The years after 1989 were characterized by a weakening of the role of spatial planning. This was accompanied with the idea that private market initiative will replace it. In parallel some attempts were also made to apply instruments of planning from the socialist era in the new conditions, but it was not feasible due to the completely new political and economic reality. The urban planning research process, that had been established so far, practically disappeared as considered for no longer useful. The leading state and municipal institutes in this field were closed—some of them like "Sofproekt - OGP" in Sofia only for a couple of years, but others were no longer restored. Education in the field of urban planning had to survive in a harsh environment and overall loss of interest towards specializations in the field.

During the rule of the socialist government, there was a lack of democratic process involving citizenship in the urban debate and decision-making. This was a serious prerequisite for this deficit to be transferred also in the next political period.

Stakeholders on the public scene had no culture and experience in open and transparent process. A long and difficult path started for its assertion. State-dominated urban solutions during the period of socialism were replaced by those on the other extreme dictated by the private capital agenda, and in both cases the public participation was neglected. The advocates of participation started to seek for its place more and more urgently, although the public interest broke down into multiple, often contradictory interests (Dimitrova 1999). Public authorities faced the challenge to moderate an open process of debate extremely unprepared, and often with a disrespectful attitude towards it.

Around the EU accession and today Bulgaria's land is under the pressure of rude real estate speculation and commodification. This is intensive at more attractive locations around commons such as coastlines, foothills of mountains and other public assets or areas with better environmental quality. Unfortunately, the speculation itself and the controversial organization of private investments maximizing short-term profits lead to great compromise in basic spatial planning and design principles and to the inefficient and inequitable development. Sofia has plenty examples to offer, especially at its more attractive south-east part adjacent to the Vitosha mountain and the contrasting more unattractive north-west part adjacent to the transit roads, old industrial estates and warehouses. After extracting the high value of land in the south-east, the investors which are underwriters and beneficiaries with their lackeys of power (Marcuse 2009) being private consultants, public administrators and decision makers redirect major public resources and spending to enhance these compromised developments. In the context of scarcity and shortage of public resources, this is at the expense of other urban areas which suffer under-development of public assets and public space more than a quarter century. Such internal and external peripheries become less and less attractive and now are under decay which perpetuates the segregation and gentrification on a citywide and long-term scale.

Thus Bulgaria's cities and regions and Sofia in particular experience such challenges that raise the question about the right to the city (Harvey 2008) as who's and what right. In the last 5 years, this question started to climb higher on the agenda of the Bulgarian society and among urban actors. Media, political groups and the citizens raised their attention in the urban domain. This is due to increasing number of civil, expert and academic groups confronting, resisting or trying to refine city governance and management, planning and urban design, architecture and investment frameworks. Today watchdogs and advocacy groups in a harsh political environment amidst media on demand, black PR, provocateurs, clowns and mimicking organizations taking part in a false moral masquerade games are campaigning and struggling for certain values and rights, rules and standards about the quality of the urban environment and the spatial development processes. Some of the early advocates for public interests active at the turn of the EU accession and then the financial crisis (before and after) were heritage protection groups (Sofia for Us 2007), environmentalists (Let Nature Remain in Bulgaria 2012; ZaZemiata 2018) as well as culturological researchers (Dichev 2010). Then a combination of young but critical architects and urbanists (Transformatori Association 2018; Grupa Grad 2013; Whata 2018), urban environmentalists (EkoObshtnost 2014), as well as public transportation lovers (Grazhdanska Initsiativa Za Obshtestven i Relsov Transport 2015; Spasi Sofia 2015) appeared along with new social action researchers (Venkov 2014; Yanchev 2014; Dinev 2016). Recently many of these actors joined forces and tried to scale up their advocacy and even to introduce deeper systematic change (ABURP 2016a; Sofia Green Capital 2018; Viziya za Sofia 2018). Some of the last developments and actions raise further questions about the relations between political power, administrative bureaucracy and civic organizations. Part of the new actors has been more successful in the process of self-empowerment. Others are at the edge of interacting and flirting with the more powerful of the day. The evaluation of outcomes from these struggles needs more distance in time.

Although the local authorities enriched their experience in management of EU funded projects, there is still a lack of administrative capacity, knowledge and sometimes will to achieve sustainability goals in spatial development. Most of the municipalities and their communities have a long way to specify their values, visions, priorities and approaches in order to use effectively the new planning instruments, especially in a more transparent way with more satisfactory involvement of their citizens. Meanwhile authorities started to realize the importance of proactive approaches besides defensive ones. Except the traditional judicial battles, partnerships with organizations and groups of well-educated and experienced specialists out of the status quo are stated as very important for improvement of the efficiency and equity of the public policy. It is because of their demonstrated potential to offer creative approaches, instruments and mechanisms for reasonable and balanced spatial development policy, taking into account the various interests. At the same time, it is a tactic to flirt and calm down the more radical critiques and questions that ruin the image of the authorities, especially around elections.

However, the potential inclusion of professional organizations is disputable because of the varying degrees of loyalty to tight economic circles or to the loose public in the treatment of important societal affairs. One of the traditional partners of state and local authorities in the planning process have been the professional organizations such as the Union of Architects in Bulgaria, founded decades ago. Their role in most cases has been to validate proposed solutions, slicked to already established practices and attitudes with the exception of architectural professional clubs and small active groups. In recent years, the scope of partners and actors has expanded with new active civil organizations and coalitions that often have rich expertise in certain urban planning and development fields. This is one of the new factors that makes the local urban agenda more contested and focuses it on issues that have been neglected or kept fragmented for years. Some of these issues had strong contribution for the spatial imbalance that persists today in Bulgarian cities.

Around the beginning of the twenty-first century, Bulgaria turned its face towards world's contemporary planning practices, experiences and knowledge. New programme in urbanism (urban planning) was introduced in the Architectural faculty of the University of Architecture, Civil Engineering and Geodesy in Sofia in the year 2002 after a long process of cooperation with European schools of planning. The programme was designed in an innovative for that time approach and with sensitivity for the European process. The accession of Bulgaria into the

EU along with other Eastern European countries was related with many reforms and transposition of EU legislation. After becoming a member of the EU in 2007 and especially before the current programming period new planning instruments and the integrated planning and the integrated territorial investment approach found their place in the national framework and in the overall spatial planning system. New ideas and actors appeared altogether with completely new challenges and possibilities.

The work of urbanists is regulated by an Act for the chambers of architects and engineers in the investment design (National Assembly of the Republic of Bulgaria 2003–2016). The activities falling under this act include the spatial planning along with the investment design. The Chamber of Architects in Bulgaria is the official representative of the professional interests of architects, landscape architects and urbanists. The inclusion of the urbanists in the Chamber of Architects was under harsh debates in the newly shaping organization. The urbanists are the smallest and youngest minority group that is misrepresented in the bodies of the Chamber. Due to this fact, there hasn't been much uptake in spatial planning matters in the work of the Chamber. ABURP is a non-governmental organization established in 2010 and working for the private interest of its members—the urbanists (urban and regional planners) but also for the improvement of the profession's image as capable to support public interests beyond the ones of the narrow plutarchy.

The controversial sometimes welcoming and other times hostile environment for the realization of planners was in the hands of two generations of "gradostroiteli" (predominantly architects specialized in territorial and urban planning and design). The first one was educated, trained and well realized during the socialist period. The second generation was educated at the end of the socialist period and was mostly realized during the Transition period. A third generation of architects interested and involved in corporate urban planning and design emerged before the crisis during the construction boom and a fourth generation that consists mostly of urbanists and a critical group of architects experienced as urban activists came out of the university incubator during the crisis and started with provocative tactical urbanism, action research and creative resistance approaches.

The ABURP has taken into consideration the local context and tried to apply the transactive planning approach along with the concepts for multilevel metropolitan governance and institutional thickness in its action research on planning and decision-making about the urban environment in Sofia. The need for enrichment of the process characterized by excessive centralization and deficits in the dialogue between individual actors has been reached and realized through a series of actions and arenas. The following lines tell the story of two interconnected projects of ABURP funded by Stolichna obshtina (Municipality of Sofia) through a programme called "Evropa". The activities included studies and elaboration of proposals for improvement of formal and informal processes, enrichment of the set of instruments for objective and subjective study of the quality of the urban environment and the facilitation of deeper dialogue between stakeholders.

Approach of the ABURP for Its Inclusion in the Process of Defining the Policy Framework for Urban Spatial Development

For two consecutive years, ABURP demonstrated in two different but interrelated projects innovative tools and mechanisms for better quality of the urban development processes. In 2016, a project "Model of shared development of liveable urban environment in the city of Sofia—"Tsar Ivan Asen II" str. and "Yavorov" residential quarter", shortly called "Share the Neighbourhood" was implemented (ABURP 2016b). In the next year, another project called "Establishment of a Standard for evaluation and monitoring of the quality of the urban environment" (in short, "Set an Urban Standard") was elaborated (ABURP 2017). Specific for both projects is the emphasis on the state of the urban environment and its future change as well as the involvement of a wide range of stakeholders in the process of planning and governance. These projects come as a result of two recognized needs: first—for new methods of thorough assessment of the urban environment, which can be applied at the lowest and closest to citizens level, and second—the need to strengthen the capacity of local communities and authorities in their active and adequate involvement in the planning process.

The ambition of the two projects is to establish an appropriate environment—both in communicational and institutional aspects where the dialogue between the widest possible range of stakeholders is facilitated by clear and impartial methods of monitoring and evaluation of the urban environment. Establishment of clear roles among the participants, based on rich, transparent and reliable data is an important step for more objectivity and fairness in decision-making at different levels—from the neighbourhood to the citywide.

Projects Goals, Partners and Territorial Scope

These two projects of ABURP share interrelated purposes, but differ significantly in their specific objectives, activities and results. "Share the Neighbourhood" was aimed to study, express and outline the identity and the uniqueness of an emblematic for Sofia space taking into account the needs of the local community. It creates an innovative analytical and well applicable cost-efficient model for development of the urban environment with broad citizen participation. The project is focused on drawing up principles, spatial design guidelines and action plan for neighbourhood development in cooperation with residents, businesses and local authorities (ABURP 2016b).

The second project, "Set an Urban Standard", creates a mechanism for monitoring and evaluation of the quality of the urban environment, which could become a more universal tool for city planning, design and governance in the context of Sofia and beyond. It aims to support more transparent and objective process. New opportunities to enrich the palette of actors and their roles in the communication are introduced, thus filling empty spaces in the institutional framework. The two projects are thematically related, where "Set an Urban Standard" is a logical sequel and upgrade of "Share the Neighbourhood" (ABURP 2017).

Trying to identify the appropriate partners, ABURP contacted several different actors: local authorities—"Sredets" district as an administrative subdivision of Sofia Municipality; "University of Architecture, Civil Engineering and Geodesy" (UACEG) as an academic institution with the capacity and experience in research; civic and student's organizations, active in the field of urbanism—IGRAD, "Transformatori" Association, Association of Students of Urbanism in Bulgaria (ASUB).

The implementation of the elaborated tools and mechanisms in both projects is directly linked with areas which are small part of the diverse spatial structure of Sofia. Both of them are located in "Sredets" district of Sofia Municipality—the first covers a residential quarter that originated in the early twentieth century, and the second one is a fragment of the city centre (spatially developing from the late nineteenth century on). They are distinctive places in the city, where concentration of diverse human activities, cultural and historical heritage can be found. "Sredets" district is characterized with active communities and variety of complex challenges.

"Share the Neighbourhood" explores a mixed-use zone with surface of 44 ha and population of 6620 inhabitants. The area combines dwellings with commercial activity, daily service, education and business activities. "Set an Urban Standard" is focused on smaller area which includes the emblematic for the city of Sofia historical garden "Kristal" and the surrounding blocks, where again combination of public and residential functions can be found—it is one of the most active places in Sofia with citywide significance.

Other territorial scopes have been envisaged and systemized in the communication of the two demonstration projects and beyond them. The vital neighbourhoods in the city centre are complemented by the old settlements around the historical centre which were further urbanized, the densely inhabited areas part of the large housing estates, along with the internal and external peripheries with green field developments, the segregated Roma neighbourhoods, the "elite" closed neighbourhoods and plenty of other variations and combinations. The extension of the experimental approaches of the two projects has been considered crucial in terms of the uptake of the innovations they introduce. Still there are many constraints for wider adoption of the experience related with the lack of dedicated national or local urban policies and funding for urban innovation, community-led development and area based regeneration in more critical liveable or unliveable urban neighbourhoods and environment under the pressure of various opportunities and threads.

"Share the Neighbourhood" Project: Model of Shared Development of Liveable Urban Environment in the City of Sofia—"Tsar Ivan Asen II" and "Yavorov" Residential Quarter

"Share the Neighbourhood" included six interrelated phases. The first of them was the conduction of desk research about the urban development and the spatial planning framework of the area. It was followed by field studies and mapping of the physical conditions in the area. The next two steps were public involvement (through an exhibition, focus groups, workshops, public debate) and collaboration between stakeholders. Then spatial development guidelines and shared action plan have been elaborated and communicated on the basis of three scenarios for the future development of the major street that is a public spline in the neighbourhood and the higher order urban plans and their impact on the landscape. The last step was sharing of the model (Fig. 13.1) among political representatives, administration, businesses and local communities.

The model was perceived as well applicable in different neighbourhoods, both in Sofia and in other cities. Its dissemination among researchers, activists, professionals and decision makers received admirations but it is still not implemented in the local planning system as an informal approach. It helps to localize the planning provisions and decisions of the differing instruments and sectoral agendas and to articulate them among the local stakeholders in order to have more coherent framework and well-informed actions for development or protection.

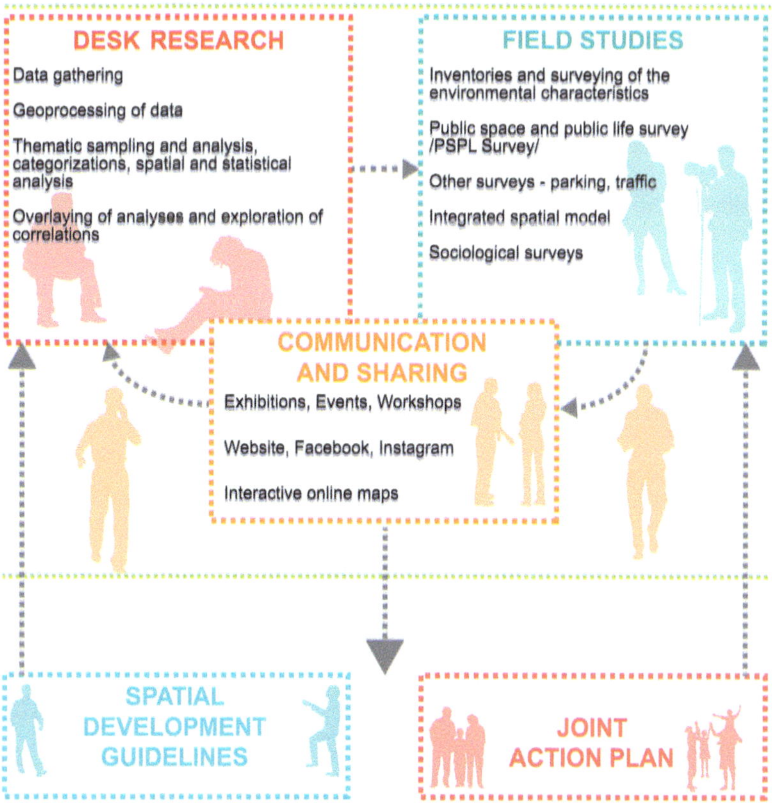

Fig. 13.1 The Model of Shared development at the neighbourhood level

"Set an Urban Standard" Project: Establishment of a Standard for Evaluation and Monitoring of the Quality of the Urban Environment

The "Set an Urban Standard" project develops a mechanism for assessment of the urban environment, including several thematic areas—healthy, safe, arranged, liveable and distinctive environment. The experience of European and World methodological standards and practices was studied (Fig. 13.2). The process of elaboration also included active consultation with experts, administrators and urban activists through interviews, workshops and consultations. The standard was tested in the chosen area through desk and field studies, thematic mapping and web-based GIS application. The project team held consultations with stakeholders on possible further development of the standard and its wider application in practice.

The proposed taxonomic framework of themes, indicators and metrics provides ground for filling of informational gaps, improvement of the agenda and accountability in the decision-making. It can bring improvements in the governance process, of the urban strategic planning and design process and of the daily management process related to the urban environment. That is considered to be crucial for the efficiency and equity of public interventions in the city space or for the balance between large public and private investments and small-scale improvements in the various urban regeneration areas or areas under urbanization in the city of Sofia and

Fig. 13.2 The standard illustrated by its five thematic areas of monitoring and evaluation

the municipality. Some of the project results were directed towards the integration of the optional work around the standard and the set of indicators part of the Integrated Plan for Urban Regeneration and Development (IPURD) of the city being partially implemented from the year 2013 on. The possible distribution of responsibilities among specific units in the municipal administration as well as additional resources and the position of the neighbourhood mancipple were outlined and discussed. Still alternative approaches (as civic and business platform) for implementation of the standard are being sought in parallel to the fragmented communication with the slowly opening and reclusive structures of the local authority which are not well integrated.

Involved Actors, Planners Inclusion and Social Learning

The actors involved during the project activities were diverse. In "Share the Neighbourhood" project, the accent was more on professionals, students, local residents, activists and businesses. In "Set and Urban Standard", the accent was put on different professionals which are part of the metropolitan municipality, national competent organs on regional level, citywide activists, urban design consultancies, major political groups and media interested in the urban development topics. Their inclusion was made possible through forums and discussions, digital social networks and public events. Specific substance matters were raised around the two demonstration areas.

Both projects "Share the Neighbourhood" and "Set an Urban Standard" were implemented with the contribution of many ABURP members and the partner organizations as well as the volunteers who took part in the activities. The major contribution of many urban planners was crucial for the successful realization of the two projects with their modest budgets, namely Angel Burov, Irina Mutafchijska, Silvia Chakarova, Nurhan Redzheb, Despina Kaneva, Vasil Madzhirski, Maya Miteva, Mila Yolova, Magdalena Kircheva, Galina Aleksova, Maria Ivanova, Beata Tsoneva, Elena Ivanova, Eli Natchkova and other contributors. The support given by the urban sociologist Nevena Germanova and the architects Marina Mateva (also landscape architect and cultural heritage expert), Elitsa Dimitrova and other technical and social science experts was invaluable.

The professional and social learning environment that the two projects provided was maybe the most important long-term outcome. Many young professionals, students and volunteers were included in an open process where inspiration, methods and techniques are being combined and shaped for the purpose of the project activities. Through the interface and exchange between keepers of different expert and experiential knowledge we see the development of a more coherent and meaningful professional language in and around the association and the urban planners group. This in turn helps in spreading important massages towards the wider public and stakeholders in other processes.

Continuity After the Two Demonstration Projects

Both projects managed to achieve the results set within their short timeframe and budget. However, they open the door for a long-term development of the themes, processes and dialogue between stakeholders. The actions, structured on the basis of the two projects that are or may be undertaken from now on, include at least the following:

- "Share the Neighbourhood"—consultation of local authorities on implementation of the model in other districts of Sofia Municipality; implementation of studies, preparation of guidelines and action plans with communities or neighbourhoods and their acceptance by local authorities and other stakeholders through the relevant agenda, responsibilities and competences; continuous observations for the spatial development of the emblematic place and its promotion as good urban form in comparison to other newly built and reconstructed areas in the city; advocacy for envisaged priority public spaces and public works in the neighbourhood.
- "Set an Urban Standard"—creation of supplemented and updated versions of the standard; integration with planning tools developed by Sofia Municipality (e.g. "Vision for Sofia"); integration into the work of municipal enterprises ("Sofproekt", "Sofia Proeкt"), as well as into the regulative framework (municipal ordinances for the urban environment under consideration); defining the criteria for sustainable and resilient development of neighbourhoods and communities in the local context; recommendations for data collection and processing; open platform for involvement of stakeholders; practical trainings of administrators, experts, etc.; integration of the standard into local community activity through the idea for "neighbourhood manciple" that can be the figure or organization bridging knowledge and organizing communication channels.

13.5 Discussion: Professional Association as Public Actors or Actors in the Name of Public Interests

The key study made a glimpse at two recent demonstration projects of the ABURP which raise many questions about:

- The role of the professional associations and their impact in society
- The strategic planning in action
- The models for shared development
- The urban knowledge and standardization of monitoring and evaluation
- The institutionalization of planning, communication and mediation
- The power and the rationalization in spatial development policy formulation and implementation

The Role of the Professional Organization in the Demonstration Projects and Beyond

Collaboration between stakeholders can be carried out by two basic approaches—the top-down and/or bottom-up. The second one of them is relatively new for the Bulgarian practice and organizations like ABURP can play an important role for its introduction. This comes as a consequence of two circumstances—first of them is that the organization bears an innovative expertise, shared among other stakeholders through successfully implemented projects, and second—it takes an independent role as an NGO. This creates an opportunity for ABURP to become a moderator in urban development debate. An evidence for this is the recognition and relatively high appreciation of its projects both from citizens and local authorities. It is interesting to mention that the ABURP has received critique during a parallel process for actualization of the local act about the consultative process in Sofia as acting deviant from the interest of the typical private urban designers and architectural consultancies. That can be considered as a good sign that it is working for more common interests first than the comfort of consultancies and authorities.

The contribution of ABURP as a professional organization in development and consolidation of rich and transactive process of planning and decision-making through both projects and resulting opportunities for future collaboration can be summarized in several key directions. The first of them is the expert support, provided by the developed tools for research, monitoring and assessment of the urban environment. The second is improvement of the communication between the various stakeholders—administration, other professionals, local communities, urban activists, businesses. It includes sharing of ideas for integrating tools into the urban planning and management system, involving more actors in the processes, allocation of responsibilities and their deployment in the day-to-day activities.

The results achieved in these two projects can be a basis for initiatives for changes in legislation on municipal level as well as introduction of new management practices in the local administration or improvement of the existing ones. Through the application of the model and the standard in real urban cases, professional organizations can be an appropriate mediator in the dialogue between stakeholders. This can be a step further to provide balance between large and focused public and private investments and small-scale improvements benefitting diverse groups.

The demonstration projects of the ABURP are considered to be efforts for promotion of more democratic, transparent and participatory culture in the spatial planning and development process at the metropolitan level and beneath. Certainly, the focus was more on the increase of the appetite for information and communication about the spatial development than on the turnabout in power asymmetries. But still the efforts are aimed at more balanced approach and clearer definition of priorities in order to make bridge between the political promises and the real distribution of public resources and development rights through more democratic mechanisms.

By being aware of the potential that the demonstrated models and tools provide, ABURP members participate in public forums, working groups and initiative committees, where they can influence other actors on their application in the real planning environment. As a result of the active participation, the organization becomes a recognizable partner, both by local authorities and civil communities in Sofia. This is also related with the ability to develop and demonstrate innovative approaches in the local contexts. Such innovations include: involvement of stakeholders during the entire planning process—from first stages of research and problem investigation to consideration of solution alternatives; detailed monitoring and mapping of the environment (using GIS); enabling a complex, more objective assessment. Further involvement of the organization members in the practice as researchers, consultants and administrators makes higher the possibility for transfer of the acquired experience and its inclusion in the practice.

One of the merits of the ABURP projects is its connection to the UACEG programme in Urbanism. The inclusion of the students as participants and volunteers in project activities is an example and proof of the learning by doing and experiencing approach. The efforts of the ABURP try to establish further an urban culture both deep grass-roots and canopy farsighted around the tree of problems and solutions for the environment of Sofia. This is a step for emersion out of the profile of selfish expert groups to benevolent professional coalitions acting about the more common interests.

Relation Between Key Issues of Spatial Development Policies Formulation and Implementation and the ABURP Projects

- *Transactive planning and transaction costs*: Model, tools and mechanisms proposed in these two projects, as well as their demonstration in a real environment, support improvement of the process of transactive planning, which includes the widest possible range of stakeholders—citizens of the respective territory, administration and businesses. The ambition is for this process to be as clear and structured as possible in terms of involving the relevant actors. Creating a favourable environment for exchange of opinions, visions and ideas also has a direct impact on the costs of organization—it has the potential to save financial and time resources in the long term.
- *Institutional thickness or thinness*: The model with its accompanying mechanisms and proposals creates a prerequisite for filling institutional gaps. This is done by "appointing" certain "actors" whose role is deemed necessary but is currently absent. The enrichment of the institutional system aims to break the current centralization, which on the one hand concentrates too much power even at expert level, and on the other hand does not offer to citizens accessible communication at the level closest to them. Current deficits determine the existence of an *institutional thinness*. Often this leads to opacity and work in private interest with a questionable, sometimes frankly contradictory public benefits. Achieving institutional thickness can lead to overcoming the current challenges.

- *Multilevel (multi-scalar) metropolitan governance*: By filling the institutional gaps and achieving the desired thickness, the interaction between the actors at different levels of governance will be developed. It is of particular importance to integrate an additional level of sub-district administrative action that has the potential to address the day-to-day problems of city users on the urban agenda. The inclusion of a wider range of experts at different levels contributes to pluralism which enriches the dialogue between stakeholders and to the generation of ideas. Improved engagement and interaction between actors, positioned in a clear communication environment, is a prerequisite for overcoming the extreme subjectivity in decision-making. Such subjectivity leads to spatial imbalances between districts and neighbourhoods in the city. Besides the realization of the large municipal infrastructure projects, the interaction process should bring the important projects and initiatives to the lowest neighbourhood level as they remain neglected over the years.
- *Network society, space of flows and space of places*: Local communities are the ones that are most sensitive to the conditions and ongoing processes in the neighbourhood or in the district that they inhabit. Often, they face challenges driven by investment initiatives, aiming to transform their place of residence. Such initiatives may stem from both private and public intentions. In both cases there is no guarantee that local needs will be met and recognized by external actors. In the case of private investment, there is a risk that the initiator will try to gain benefits of the expense of values that are important for the community. In case of public investments, there is a risk that local authorities can sacrifice such values in order to solve citywide problems or because of careless planning approach with a lack of communication with the locals in early stages. But there are new actors in the process that have increasing influence on the urban stage of Sofia—the active civil society organizations that may not be representatives of a local community, but they are watching for the quality of the city's environment as a whole. This type of actors has the potential to help building or retaining the local community that would otherwise be too vulnerable to the turbulent movement of external capitals.
- *Self-reliance and resilience*: Provided environment for communication, problem-solving and exchange of ideas is an important foundation for a complete institutional framework that has the capacity to effectively organize the desired measures and the resulting costs. This can contribute to the adaptability and resilience of neighbourhoods and their communities to challenges arising from transformation in the external environment.
- *Liveability and living urban environment*: The enriched pluralistic process has better potential than the conventional centralized model to point out more clearly the challenges of different city districts and neighbourhoods. Some of these neighbourhoods are characterized by lost or never possessed vitality. There are areas situated close to large transport corridors with very intensive individual motorized traffic. There, a healthy, safe and vibrant environment is sacrificed in the name of a citywide effect, with dubious benefits for the sustainability of

urban mobility. In other cases, in currently built neighbourhoods acute deficits in the social and transport infrastructure can be seen. Environmental measures, public space and community infrastructure, tactical urbanism, urban acupuncture, creative resistance art and the like can significantly improve the quality of the urban environment and local public life.

New Role of Urban Planners and Resulting Innovations

Contemporary planning processes not only in Bulgaria but also in Central and Eastern Europe as a whole are marked by the influence of neo-liberal market phenomena. This inevitably puts pressure on city decisions for urban development, which gives preference to private corporate interests, many times at the expense of the public ones. But public interests themselves cannot always be easily identified. The need for access to public goods and services that the urban environment provides at different levels—from the neighbourhood to metropolitan area, may be controversial for individual groups of users. And here one of the most important roles of the urbanists is emerging—through their expertise and overall view of the city, they can help for identification of the public interest in its different scales. The public interest itself can go beyond the local and the metropolitan level and meet regional, national and even trans-border needs. Identification is only the first step and it is a basis for subsequent assertiveness of the public interest. Urbanists have the most combinative position on the complex view of the build environment, its systems and ongoing processes by default. The profession has the task to focus on objectivity and search for the truth, but it also can enable serious manipulations. This defines the ethics as a leading challenge to professionalism.

The knowledge and skills that professionals use in identification and protection of the public interest determine their role as initiators and moderators of the urban debate. This is the area where innovations can be searched. In the local context of Sofia, the active involvement of local stakeholders in the entire planning process and the integration of their needs in the final result can be defined as innovation. Another innovation is the application of research methods for detailed monitoring and assessment of the environment, which helps for more objective definition of problems and identification of public interest.

The great challenge in the current situation is to turn these innovations into routine practice. Professional organizations should assert directly this need in front of the local authorities, but also communicate it with the active community in order to create joint efforts influencing the decision makers. All of this is a prerequisite for making valuable city decisions that must be motivated by a good assessment of impact on the complex environment. The responsibility of urbanists here is to support a process in which communities feel more empathetic towards major choices for transition and adaptive cycles on the different levels.

References

Alexander C (1977) A pattern language: towns, buildings, construction. Oxford University Press, New York

Allmendinger P (2002) Planning theory. Palgrave, Basingstoke

Amin A, Thrift N (1995) Globalisation and institutional thickness. In: Healey P, Cameron S, Davoudi S, Graham C, Madanipour A (eds) Managing cities: the new urban context. Wiley, Chichester, pp 91–108

Appleyard D, Gerson MS, Lintell M (1981) Liveable streets. University of California Press, London

Arnstein SR (1969) A ladder of citizen participation. Journal of the American Planning Association 35(4):216–224

Association of Bulgarian Urban and Regional Planners (ABURP) (2016a) Urban Development Forum. http://bgplanning.org/projects/forum-za-gradsko-razvitie/. Accessed 12.02.2018

Association of Bulgarian Urban and Regional Planners (ABURP) (2016b) Model of shared development of vital urban environment in city of Sofia – "Tsar InavAsen II" str. and "Yavorov" residential quarter. http://bgplanning.org/projects/spodeli-kvartala/. Accessed 12.02.2018

Association of Bulgarian Urban and Regional Planners (ABURP) (2017) Establishment of standard for evaluation and monitoring of quality of the urban environment. http://bgplanning.org/projects/gradski-standart/. Accessed 12.02.2018

Campbell S (2003) Green cities, growing cities, just cities? Urban planning and the contradictions of sustainable development. In: Campbell S, Fainstein S (eds) Readings in planning theory, 2nd edn. Blackwell, Oxford

Castells M (2008) Space of flows, space of places: materials for a theory of urbanism in the information age. In: Haas T (ed) New urbanism and beyond: designing cities for the future. Rizzoli International Publications, New York

Centre for Study of Democracy (CSD) (2016) State capture unplugged: Countering administrative and political corruption in Bulgaria: http://www.csd.bg/artShow.php?id=17723. Accessed 12.02.2018

COST Action C20 (2009) Urban Knowledge Arenas: re-thinking urban knowledge and innovation. Final report of COST action C20. http://www.cost.eu/module/download/7518. Accessed 12.02.2018

Davoudi S, Shaw K, Haider LJ, Quinlan AE, Peterson GD, Wilkinson C, Fünfgeld H, McEvoy D, Porter L (2012) Resilience: a bridging. concept or a dead end? Planning Theory & Practice 13(2):299–333

Dempster MBL (1998) A self-organizing systems perspective on planning for sustainability. A thesis presented to the University of Waterloo in fulfilment of the thesis requirement for the degree of Master of Environmental Studies in Planning, Waterloo, ON. http://www.bethd.ca/pubs/mesthe.pdf. Accessed 12.02.2018

Dichev I (2010) Sluchvaniya na Grada Otdolu [Grassroots approach in city implementations]. https://www.seminar-bg.eu/spisanie-seminar-bg/broy2/item/240-sluchvaniya-na-grada-otdolu.html. Accessed 12.02.2018

Dimitrova E (1999) Ustoychivost na gradskoto razvitie. Gordost I predrazsadatsi v prostranstvenoto planirane na postindustrialniya grad, Nauchna konferentsiya s mezhdunarodno uchastie "Arhitekturata kato izkustvo – 50 godini arhitekturni izsledvaniya v BAN" [Sustainablity in urban development. Pride and prejudice in spatial planning of postindustrial city, Scientific conference with international participation "The Architecture as an Art – 50 years of scientific research in Bulgarian Academy of Science"], BAS, Sofia

Dinev I (2016) Studentski Grad na Studentite!" Edna Istoriya za Pravoto varhu Grada" [Student's City to the Students! A story about the right to the city]. http://dversia.net/1102/pravoto-varhu-grada/. Accessed 12.02.2018

EC (2017) Report from the Commission to the European Parliament and the Council. On Progress in Bulgaria under the Co-operation and Verification Mechanism. COM (2017) 750 final. https://ec.europa.eu/info/sites/info/files/comm-2017-750_en_0.pdf. Accessed 12.02.2018
EkoObshtnost (2014) Idei za po-efektiven Dialog Mezhdu Grazhdani I Stolichna Obshtina [Ideas for more effective dialogue between citizens and Sofia Municipality]. http://bepf-bg.org/идеи-за-по-ефективен-диалог-между-граж-2. Accessed 12.02.2018
EU (2016) Report from the Commission to the Council on the Urban Agenda for the EU. COM(2017) 657 final:https://ec.europa.eu/futurium/en/system/files/ged/celex3a52017d-c06573aen3atxt.pdf. Accessed 12.02.2018
Flyvbjerg B (2002) Planning and foucault: in search of the dark side of planning theory. In: Allmendinger P, Tewdwr-Jones M (eds) Planning futures: new directions for planning theory. Routledge, London, pp 44–62
Forester J (2007) Planning in the Face of Power, Journal of the American Planning Association, 48:1, 67–80
Friedmann J (1973) Retracking America: a theory of transactive planning. Doubleday/Anchor, Garden City
Friedmann J (1987) Planning in the public domain: from knowledge to action. Princeton University Press, Princeton
Friedmann J (2003) Why do planning theory? Planning Theory 2003; 2; 7. (2011) Insurgencies: essays in planning theory. Routledge, New York
Grazhdanska Initsiativa Za Obshtestven I Relsov Transport (2015) Monthly Archives: November 2015: https://obshtestven.wordpress.com/2015/11/. Accessed 12.02.2018
Grupa Grad (2013) Kritika [Ctiticism]: http://www.grupagrad.com/. Accessed 12.02.2018
Gunderson LH, Holling CS (2003) Panarchy: understanding transformations in human and natural systems. Island Press, Washington, DC
Hacking N, Flynn A (2017) Networks, power and knowledge in the planning system: a case study of energy from waste. Progress in Planning 113:1–37
Harvey D (1985) The urbanization of capital. Johns Hopkins University Press, Baltimore, p xvii, 239
Harvey D (2008) The right to the city. New Left Review II(53):23–40
Healey P (1992) Planning through debate: the communicative turn in planning theory. The Town Planning Review 62(2):143–162
Healey P (1997) Collaborative planning – shaping places in fragmented societies. Macmillan, London
Institute for Market Economics (IME) (2017) Democratic Backsliding in Bulgaria. http://ime.bg/bg/articles/otstyplenieto-na-demokraciyata-v-bylgariya/. Accessed 12.02.2018
"Let Nature Remain in Bulgaria" Coalition (2012) Aarhus convention - the new beginning for Bulgarian spatial planning in the favor of citizens. http://forthenature.org/news/2504. Accessed 2.02.2018
"Let Nature Remain in Bulgaria" Coalition (2018a) Sluchai [Cases]. http://forthenature.org/cases. Accessed 12.02.2018
"Let Nature Remain in Bulgaria" Coalition (2018b) Uchastie za Prirodata [Participaion for Nature]. http://forthenature.org/uchastie-za-prirodata, Accessed 12.02.2018
Marcuse P (2009) From critical urban theory to the right to the city. City 13(2–3):185–197
National Assembly of the Republic of Bulgaria (2003–2016) Zakon za Kamarite na Architektite I Inzhenerite v Investicionnoto Proektirane [Law of Chambers of Architects and Engineers in Investment Design]. State Gazzette
Piattoni S (2009) Multi-level governance: a historical and conceptual analysis. European Integration. 31(2):163–180
Regional Environmental Center for Central and Eastern Europe (REC) (1998) Doors to democracy: current trends and practices in public participation in environmental decisionmaking in Central and Eastern Europe. http://archive.rec.org/REC/Publications/PPDoors/CEE/PPDoorsCEE.pdf. Accessed 12.02.2018

Sofia for Us (2007) Tazhn ivesti [Sad news]: http://sofiazanas.blogspot.bg/2007/11/blog-post_8153.html. Accessed 12.02.2018

Sofia Green Capital (2018) Za Nas [About us]: https://sofiagreencapital.org/за-нас. Accessed 12.02.2018

Spasi Sofia (2015) Zapochvame [We start]. http://spasisofia.org/bg/започваме.html. Accessed12.02.2018

Szelenyi (1983) Urban inequalities: under state socialism. Oxford University Press, Oxford

Transformatori Association (2018) Komunarchia. https://transformatori.net/komunarhia/. Accessed 12.02.2018

UN (2016) Habitat III Policy paper 6: Urban spatial strategies: land market and segregation: http://habitat3.org/file/524217/view/572975. Accessed 12.02.2018

UNECE (2018) Bulgaria Decision V/9d. https://www.unece.org/environmental-policy/treaties/public-participation/aarhus-convention/envpptfwg/envppcc/envppccimplementation/fifth-meeting-of-the-parties-2014/bulgaria-decision-v9d.html. Accessed 12.02.2018

Venkov N (2014) Konspiratsii okolo Zhenskiya Pazar [Conspiracies about "Woman's market"]: https://www.seminar-bg.eu/spisanie-seminar-bg/broy10b/item/417-konspiracii-okolo-zhenskiya-pazar.html. Accessed 12.02.2018

Viziya za Sofia (2018) ZaViziata [About the Vision]: https://vizia.sofia.bg/sofia-vision/. Accessed 12.02.2018

Whata (2018) Whata awards. http://whata.org/awards/ Accessed 12.02.2018

Williamson O (1979) Transaction-cost economics: the governance of contractual relations. Journal of Law and Economics 22(2):233–261

Yanchev P (2014) Epizodi na Urbanistichen Aktivizam v Sofiya – Gradskite Publichni Prostranstva Mezhdu Institutsii i Grazhdani [Episodes of urbanistic activism in Sofia – city public spaces between institutions and citizens]. https://www.seminar-bg.eu/spisanie-seminar-bg/broy10b/item/420-epizodi-na-urbanistichen-aktivizam-v-sofia.html. Accessed 12.02.2018

ZaZemiata (2018) Deynosti po Upravlenie na Otpadatsite v gr. Sofiya [Waste management activities in tne City of Sofia]. http://zazemiata.org/v1/Sofija.273.0.html. Accessed 12.02.2018

Index

© Springer International Publishing AG, part of Springer Nature 2019

M. Finka et al. (eds.), *The Role of Public Sector in Local Economic and Territorial Development*, EAI/Springer Innovations in Communication and Computing, https://doi.org/10.1007/978-3-319-93575-1

MIX
Papier aus verantwortungsvollen Quellen
Paper from responsible sources
FSC® C105338
FSC
www.fsc.org

If you have any concerns about our products,
you can contact us on
ProductSafety@springernature.com

In case Publisher is established outside the EU,
the EU authorized representative is:
Springer Nature Customer Service Center GmbH
Europaplatz 3, 69115 Heidelberg, Germany

Printed by Libri Plureos GmbH
in Hamburg, Germany